Beryl

And Its Color Varieties

Contributions by:

Dmitriy Belakovskiy, John Bradshaw, Bruce Cairncross, Alexander Falster, Jesse Fisher
Lee Groat, Dermot Henry, Rupert Hochleitner, Michael Huber, Lee Van Iderstine, Mark Jacobson
Bruce Jarnot, Guanghua Liu, Peter Lyckberg, Luiz Menezes, Alfredo Petrov, Tony Potucek
Federico Pezzotta, William "Skip" Simmons, Fabio Tamagnini, Stefan Weiss and Michael Wise

Edited by:

Alexander Falster, Miranda Jarnot, Günther Neumeier, William "Skip" Simmons
Gloria Staebler, Tom Wilson and Michael Wise

Illustrations by:

Roberto Appiani, Paul Balayer, Bruce Cairncross, Frank Coffa, Emanuela Cucchiari, Kevin Dixon
Jesse Fisher, Gary Freeman, Bernie Gaboury, Walter Gabriel, Adalberto Giazotto, Lee Groat
Tino Hammid, John Haupt, Peter Huber, Terry Huizing, Jan Kanis, Dawn Kellett, Peter Lyckberg
Luiz Menezes, Günther Neumeier, Eckehard Julius Petsch, Roberto Pucci, George Rossman
Thomas Schüpbach, Jeff Scovil, Josie Scripps, Edgardo Signoretti, William "Skip" Simmons
Joel Sweet, Harold and Erica van Pelt, Stefan Weiss, Fred Wilda and Brad Wilson

Lapis International, LLC · USA

Beryl and Its Color Varieties

Friedrich August Breithaupt's Beryl Seal

Mineralogist Friedrich August Breithaupt had his seal carved into a beryl crystal. Breithaupt began his studies in 1811, during A.G. Werner's time at the Mining Academy in Freiberg, Saxony (Germany). From 1826 to 1866, he was responsible for the mineralogical education of Freiberg's miners. He was also the gemstone inspector for the mining authority. Breithaupt's legacy is 500 publications, including 46 first descriptions.

The above is a postcard of the School of Mining in Feiberg, Saxony printed for the 1991 Breithaupt Symposium. Günter Grundmann archive

Table of Contents

Right: Aquamarine Rough from Marijao, Madagascar *is sorted in a workshop in Idar-Oberstein, Germany.*

Photo Eckehard Julius Petsch

Mineralogy of Beryl

William "Skip" Simmons and Alexander Falster, pegmatite specialists from the University of New Orleans in Louisiana (USA), with the lowdown on beryl mineralogy

Beryl is one of the oldest known minerals, though the origin of the term "beryl" is unclear. It may have been derived from the Greek term "beryllos," which was used to refer to blue-green gems in general and was probably applied to several different minerals.

The Latin term "beryllus" may have been used to refer, rather than to a given species, to the crystal properties of clarity and luster. The French term "briller," meaning "to shine," as well as "brille," the German word for "glasses," are both rooted in the Latin term "beryllus."

Beryl is a beryllium aluminum cyclosilicate important to both the gem industry and to collectors. It occurs in magnificent crystals and in a wide range of beautiful colors, the more common of which have varietal names that are widely used (see page 22). These include aquamarine, red beryl, morganite, emerald, goshenite, and heliodor.

Known for millennia, the beryl varieties emerald and aquamarine are surrounded by the ancient mysticism of birthstones. Emerald, aquamarine and beryl were all named in Pliny's *Natural History* (ca 79 A.D.), though Pliny used different terms to refer to these stones. Book 37 of Pliny's work deals with the beryl varieties. In his 1950 volume *A Roman Book on Precious Stones*, Sydney Ball states with some certainty the modern minerals to which Pliny was referring in his ancient text: "Egyptian and Aethiopian smaragdus" referred to what would be known today as emerald; "beryllus" is the Latin term for beryl; "sea green beryllus" referred to aquamarine; and "chrysoberyllus" was the term for golden beryl. Pliny used the term "hyacinthozontes" for deep blue beryl and "aeroides" for pale blue beryl.

According to John Sinkankas, who literally wrote the book on beryl, with the exception of emerald, the modern designations for the beryl varieties emerged much later than did "emerald" and "beryl." The term "aquamarine" first appeared in print in Anselm Boetius de Boodt's landmark 1609 lapidary *Gemmarum et Lapidum Historia*, after which it was quickly accepted. The other varietal names came into being after the 17th century.

In spite of the valiant attempts of Pliny and the authors and scientists that followed him, it was not until the late 18th century developments in chemistry that the harder gemstones could be analyzed accurately. Until that time, terms such as "emerald" were used to refer to any hard, green stone, and numerous other mineral species were lumped together as "beryls." Likewise, without the benefit of chemical and X-ray analyses, many early scientists assumed that the different color varieties of beryl were indeed separate species.

Mineralogy and gemology made great strides in the 19th and 20th centuries, and modern analytical methods have provided a great deal of information about the structure and composition of minerals. Today, in addition to the mineral beryl, the term "beryl" applies to a group of minerals consisting of the mineral beryl and four closely related cyclosilicate species: bazzite (page 16), stoppaniite (page 20),

*An ancient gemstone, aquamarine is well suited to modern design. The Bernd Munsteiner aqua and diamond necklace **below** is in the collection of Michael Scott. van Pelt photo*

and the Beryl Group

indialite (page 10), and the newly discovered pezzottaite (page 11). The most important and widespread member of this group is the mineral beryl. In this publication, we will differentiate between the mineral and the group by using the term "beryl" to refer to the mineral and the words "beryl group" to refer to the group.

Beryl Group Composition

The formula for beryl is typically given as $Be_3Al_2\ Si_6O_{18}$. The crystal structure consists of Si_6O_{18} rings, which are linked to one another by beryllium and aluminum. The Si_6O_{18} are stacked a top one another in perfect alignment along the *c*-axis, creating channels in the crystal structure that run parallel to the *c*-axis. In most beryl group minerals, some water and cations, particularly alkalis such as sodium, potassium, cesium and rubidium, are contained within the channels. The crystal structures of beryl and pezzottaite are illustrated and discussed on page 14.

Over the years, numerous formulae have been proposed to account for the alkalis and water found in the channels. For example, Ginsberg proposed $R^{1+}Be_{3-0.5n}Al_2Si_6O_{18}\cdot pH_2O$, where R^+ represents the alkali ions; n can range from 0 to 1; and p from 0.2 to 0.8. Feklichev (1963) suggested the formula $R^+_{n-1}\ R_{2+n}\,Be_{2.5/2n}Al_2Si_6O_{18}\cdot pH_2O$, in which R^+ represents cesium, lithium, rubidium, sodium, or potassium. The lithium present in most cesium-bearing beryl has now been shown to substitute for beryllium.

A general structural formula that accounts for the chemical variation found in the beryl group can be expressed as $\mathbf{R\ X_3\ Y_2\ (T_6O_{18})\cdot pH_2O}$, with the letters R, X, Y, T and p referring to specific positions within the crystal structure of beryl. The ions that can occupy each given site are as follows:

R= Cs^{1+}, Na^{1+}, □ (in the channel site)
X= Be^{2+}, Al^{3+}, Li^{1+}, (Si^{4+})
Y=Al^{3+}, Sc^{3+}, Fe^{3+}, Mg^{2+}, (Cr^{3+}), (V^{3+}), (Fe^{2+})
T=Si^{4+}, Al^{3+}
p~0.05 - 1.8

□ *indicates a vacancy (missing ions from a site)*
Ions in ( ) indicate minor substitution

The end-member compositions of the five individual minerals of the beryl group are shown in the table on the right.

Left: *Beryl crystals from various alluvial deposits; the largest crystal is about 4 cm wide*

Collection Eckehard Petsch

Color

The mineral beryl occurs in numerous colors ranging from colorless to white, blue, green, yellow to brown, pink, and red. A number of varietal names are widely, though imprecisely, used to designate some of the more common beryl colors (see page 22).

Common beryl is generally opaque or translucent, while more rare transparent crystals are highly prized for gem rough and as collector specimens. Larger crystals may have a cloudy or mottled appearance as a result of abundant fluid inclusions or areas of alteration. Black is very

Beryl Group End-Member Compositions

Sites	(R)	(X)	(Y)	$T_6\ O_{18}$
Species				
Beryl	□	Be_3^{2+}	Al_2^{3+}	$Si_6\ O_{18}$
Bazzite	□	Be_3^{2+}	Sc_2^{3+}	$Si_6\ O_{18}$
Stoppaniite	□	Be_3^{2+}	Fe_2^{3+}	$Si_6\ O_{18}$
Pezzottaite	Cs	$Be_2^{2+}Li^{1+}$	Al_2^{3+}	$Si_6\ O_{18}$
Indialite	□	$(Al_2^{3+}Si^{4+})$	Mg_2^{2+}	$(Al_2\ Si_4)\ O_{18}$

Color zoning in beryl, as seen in the pink and blue crystal ***above*** *from Pakistan, is not uncommon. This specimen is in the collection of Herb Obodda.*

Photo Jeff Scovil

uncommonly seen in beryl and is usually due to a high density of small black inclusions. Color zoning is frequently observed.

Beryl color is responsive to heat treatment and irradiation. Heat treatment under reducing changes green or yellow beryl to blue; whereas heating blue or green beryl under oxidizing conditions changes the color to shades of yellow. Other color changes can be brought about by irradiation with gamma rays or X-rays, though not all of these color-enhancements are permanent (Sinkankas, 1981).

The cause of color is most likely the result of a complex interaction between major elements, trace elements, structural defects, and possibly natural irradiation. Nevertheless, the color in most common green beryl is caused by iron. Although the details in many cases are imperfectly understood, certain colors are thought to be related to the presence of specific chromophoric elements (see table below).

Occurrence

With the exception of the emerald and red beryl varieties (page 98), beryl forms primarily in granitic pegmatites. Most ore beryl occurs in massive pegmatite zones where crystals may reach considerable size (see box below). In fact, crystals in excess of 20 tons are known (Sinkankas, 1981).

Euhedral beryl, including goshenite, aquamarine and morganite, can be found adjacent to the core margin of the pegmatite and in

Suggested Chromophores for Beryl

Color	Chromophore	Color-generating mechanism	Reference
Green	Fe^{2+}, Fe^{3+}	Molecular orbital transition	Schmetzer, Berdesinski & Bank (1974)
Blue	Fe^{2+}	Molecular orbital transition	Wood & Nassau (1968)
Yellow and orange	Fe^{3+}	Molecular orbital transition	Wood & Nassau (1968)
Pink	Mn^{2+}, Mn^{3+}	Crystal field transition	Sinkankas (1981)
Red Utah type	Mn^{3+}	Crystal field transition	Platonov, Taran & Klyakhin (1989)
Emerald green	Cr^{3+} and/or V^{3+}	Crystal field transition	Wild & Klemm (1926)
colorless	None, no change after irradiation		Sinkankas (1981)

Below: *Rupert Hochleitner poses for scale next to a giant beryl crystal at Senhora da Assunção in Portugal*

Giant Crystals

Beryl is among the relatively few minerals that forms huge crystals. A few of the largest on record are listed below. All of these giants formed in pegmatites.

1m: Meter long crystals have been found in **Lemnäs, Finland**, where the largest known crystal is 1 meter long and 25 cm in diameter; **Ohigashi, Japan**; and in **Dronning Maud Land, Antarctica**, where meter long blue green crystals are known.

1.3m: A 1.3 meter long crystal has been recorded from the **Alger quarry in New Hampshire (USA)**. The crystal weighed 455 kilograms.

2m: The **Queen Anne mine in Zimbabwe** has produced crystals up to 2 meters in length and weighing up to 350 kilograms.

3m: Crystals 3 meters long by 1.5 meters wide have been found at **Senhora da Assunção, Portugal**. Pegmatites at **Mahajamba, Madagascar** have yielded beryl crystals to lengths of 3 meters, as have those at **Tin Mountain, South Dakota.**

4m: Beryl crystals have reached 4 meters at **Cerro Blanca, Argentina**.

8.4m: The largest beryl crystal at the **Bumpus quarry** in **Maine** was 8.4 meters long, 1.9 meters in diameter and weighed almost 24 tons.

8.7m: In **Jackalswater, South Africa**, the largest beryl crystal recorded was 8.7 x 1.5 x 1.5 meters and weighed 16 tons.

18m: The record holder for the world's largest known beryl crystal is a pegmatite at **Malakialina, Madagascar**, which produced one crystal that was 18 meters high and 3.5 meters wide.

miarolitic cavities. The size of the pocket (euhedral) beryl crystals can be respectable, sometimes exceeding 1 meter, but they do not approach the giant-sized crystals found in the massive pegmatite portions.

Morphology

Beryl crystals are hexagonal and strongly prismatic. They are commonly striated, grooved and elongated parallel to the *c*-axis, with large tapered crystals occurring in some pegmatites.

The forms most commonly seen in beryl crystals are the hexagonal prism $\{10\bar{1}0\}$ and the basal pinacoid $\{0001\}$. Modifications by dipyramids $\{11\bar{2}1\}$ or $\{10\bar{1}1\}$ and the dihexagonal prism occur rather commonly in beryl crystals, but pyramidal crystals are rare.

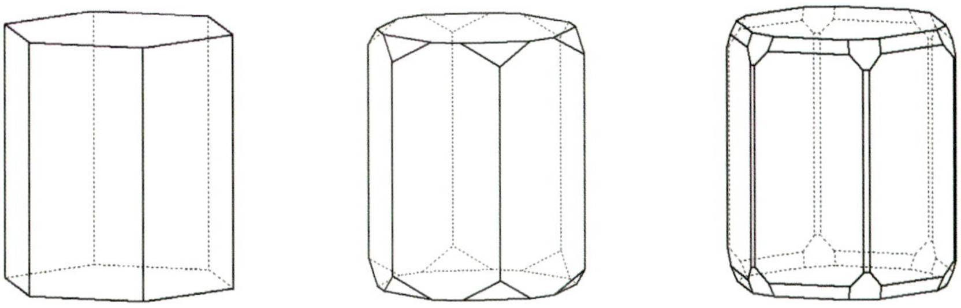

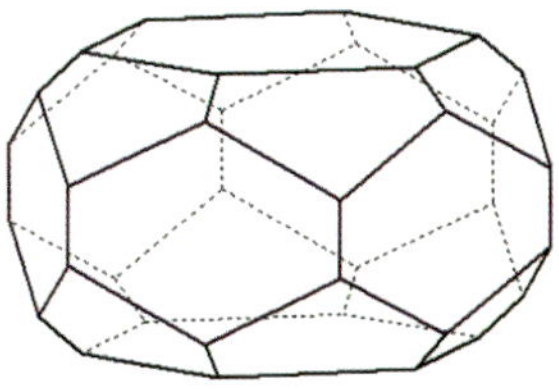

Common Beryl Morphologies

***From upper left to lower right**: pinacoids and prism; pinacoids, prism, and dipyramid $2\bar{1}\bar{1}2$; pinacoids, prism, as well as 1st and 2nd order dipyramids $2\bar{1}\bar{1}1$ and $2\bar{1}\bar{1}2$; thick tabular crystal, pinacoids, prism, dipyramids $2\bar{1}\bar{1}2$*

Aquamarine, heliodor, and goshenite are typically prismatic in habit, though both goshenite and aquamarine are sometimes found as tabular or platy crystals. Dipyramidal forms are generally small but may become more prominent in morganite, which also tends to form thick tabular crystals. Cesium-rich beryl is characteristically flattened on the basal pinacoids.

Etch pits and dissolution features are widespread, especially in gem-quality beryl from pegmatite pockets.

Tapered crystals are generally the result of strong degrees of undercooling, which may be the result of rapid cooling or of some type of chemical quench, both of which result in rapid crystallization.

*Gem pocket beryl often exhibits evidence of dissolution, as seen in the blue beryl on feldspar from the Northern Areas, Pakistan (**upper left**; collection Dudley Blauwet; photo Jeff Scovil) and the green crystal from Volodarsk-Volynskiy, Ukraine (**upper right**).*

*The crystal pictured **right** exhibits tapering typical for beryl from some pegmatites. This specimen, from Minas Gerais, Brazil, is housed in the Deutsches Edelstein Museum in Idar-Oberstein, Germany. Photo Terry Huizing*

This 9 cm high gemmy beryl crystal exhibit an interesting growth structure running parallel to the c-*axis. It is from the 60 km marker, near Sabsar, on the Gilgit to Skardu Road, Skardu district, Northern Areas, Pakistan.*

Collection Ty Lawrence; photo Kevin Dixon

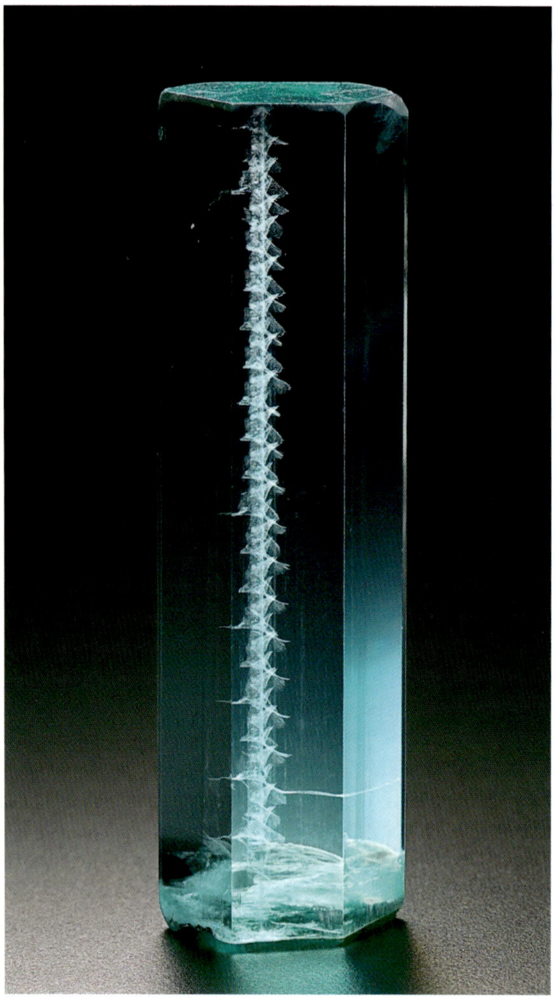

Physical Properties

The streak of beryl is white, and the luster of freshly broken beryl is glassy. Fractured surfaces are smooth or conchoidal. Beryl has an imperfect cleavage on {0001}, though the basal cleavage is somewhat more pronounced in emerald. The Mohs hardness of beryl ranges from 7 ½ to 8 and has been reported as low as 7. Density ranges from 2.63 to 2.92 g/cm^3 for beryl with elevated alkali content, especially cesium. As a point of comparison, the densities of all of the beryl group minerals are plotted on the upper right. Some beryl fluoresces white in short and long wave ultraviolet light.

At room temperature, beryl is insoluble in most mineral acids and dissolves in hydrofluoric acid only with difficulty. It is soluble in sodium or potassium hydroxide (Anthony et al. 1995). Because of its resistance to chemical attack and its strong crystalline structure, beryl is extremely durable. Remarkably unaltered crystals are found in soils and stream gravels in Brazil. On the other hand, at high temperatures and pressures, beryl breaks down relatively easily; thus, it is not uncommon to find etched or altered beryl crystals *in situ*.

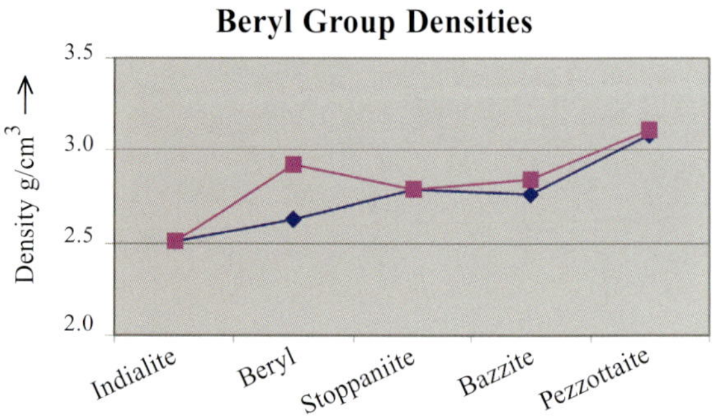

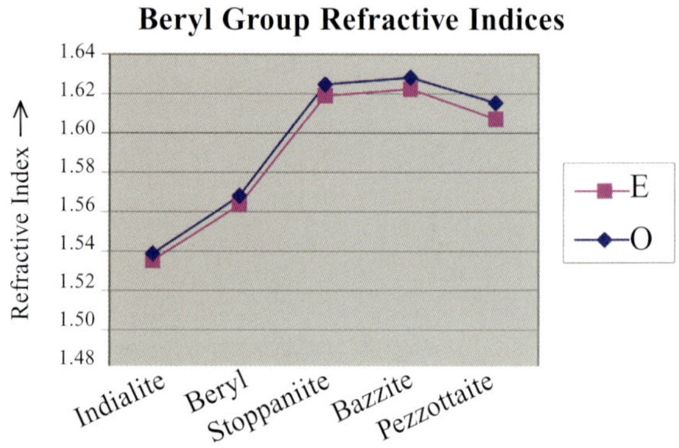

Optical Properties

Beryl is uniaxial negative. Crystals are length fast with parallel extinction. Some beryl crystals are reported to display anomalous biaxial optics with a 2V up to 17 degrees. Emerald is slightly pleochroic.

The indices of refraction have been carefully measured for beryl because of its importance in the gem industry. They are listed in the table below.

Refractive Indices for the Color Varieties of Beryl

Goshenite (low alkali):	N_O 1.568	N_E 1.564	birefringence = 0.004
Emerald (low alkali):	N_O 1.576-1.593	N_E 1.593-1.586	birefringence = 0.005-0.007
Aquamarine (low alkali):	N_O 1.575-1.585	N_E 1.570-1.580	birefringence = 0.006
Red beryl (low alkali):	N_O 1.576	N_E 1.570	birefringence = 0.006
Morganite (high alkali):	N_O 1.585-1.598	N_E 1.578-1.589	
High alkali-Fe-Mg beryl:	N_O 1.608	N_E 1.599	birefringence = 0.006-0.009 (Gaines et al. 1997)

X-ray Diffraction Lines

The following 8 strongest reflections (in 2Θ, intensities in parentheses) are characteristic for beryl: 7.98 (90), 4.60 (50), 3.99 (45), 3.254 (95), 3.015 (35), 2.867 (100), 2.523 (30), 1.9926 (20) (JCPDS #9-430). These are shown in comparison to the other species in the beryl group in the table shown at the bottom of the page.

Paragenesis

The paragenesis of beryl depends principally on its pegmatite affiliation. In less evolved beryl and beryl-columbite phosphate type pegmatites, beryl is associated with quartz, microcline, albite, muscovite, biotite, ferrocolumbite, and phosphates (such as triphylite) as well as with the common alteration and replacement products of beryl.

In more evolved pegmatite systems, such as the complex type (lepidolite, amblygonite, elbaite and danburite subtypes) beryl can be associated with microlite, more evolved members of the columbite-tantalite group, cookeite, tourmaline, lepidolite, topaz, and spessartine (Pezzotta, 2001; Černý, 1992).

In greisen deposits, beryl is associated with topaz, cassiterite, and ferberite-hübnerite; whereas in Alpine and hydrothermal veins, it is found with quartz and feldspar. In topaz rhyolites where red beryl is found, topaz, high-temperature quartz, and bixbyite are common associates of beryl.

Alteration

At high temperatures and pressures, beryl commonly alters partially or completely to several secondary mineral

***Above:* Beryl, topaz, albite, and quartz**; *15 cm from the Shigar Valley, Skardu district, Northern Areas, Pakistan Because it is associated with topaz, this beryl crystal is probably the product of a more evolved pegmatite.*

Collection Wayne Sorensen; photo Jeff Scovil

Below*: The seven strongest X-ray powder diffraction lines for the minerals of the beryl group in order of decreasing intensity; the units are d values with the intensities in parentheses*

Species	d (I)	d (I)	d (I)	d (I)	d (I)	d (I)	d (I)
Beryl	2.867 (100)	3.254 (95)	7.98 (90)	4.60 (50)	3.99 (45)	3.015 (35)	2.523 (30)
Bazzite	2.94 (100)	3.29 (100)	1.646 (80)	8.2 (70)	2.56 (50)	2.28 (50)	1.795 (50)
Stoppaniite	3.278 (100)	8.12 (65)	2.903 (64)	4.00 (39)	2.553 (18)	1.752 (18)	
Indialite	8.48 (100)	3.027 (85)	3.138 (65)	3.379 (55)	4.094 (50)	4.89 (30)	1.688 (30)
Pezzottaite	3.271 (100)	2.871 (50)	3.027 (40)	3.019 (30)	2.215 (14)	1.636 (14)	2.229 (12)

species. Low pH fluids (pH~2-3) result in quartz as the dominant alteration product. A pH of 4-5 produces bertrandite, euclase, or phenakite. A near neutral pH~7 will produce bertrandite or bavenite whereas a pH of ~8-9 results in bavenite, milarite, or bityite. More alkalic fluids of a pH ~10-11 will produce epididymite or eudidymite (Černý, 2002).

Of these secondary species, bertrandite is the most common. In the presence of metasomatic phosphate-bearing fluids, Be-phosphates such as uralolite, hydroxylherderite-herderite, hurlbutite, gainesite, väyrynenite, beryllonite, roscherite, moraesite, selwynite, faheyite, fransoletite, greifensteinite, mccrillisite, ehrleite, pahasapaite, parafransoletite, tiptopite, weinebeneite, and zanazziite may result (Burt, 1975; Simmons et al. 2003). At high temperatures and pressure, beryl becomes unstable and breaks down to chrysoberyl, phenakite, and quartz (Burt, 1978).

Locations

Locations for beryl are numerous and this volume highlights some of the most important including a number of Brazilian states (page 84); San Diego County, California (page 32) in the United States; Canada (page 40); pegmatites in Russia and the Ukraine (page 49); as well as numerous localities in southern Africa (page 78). More recent is the large production of wonderful aquamarine from Afghanistan, and China (page 64) is an emerging source for many fine mineral and gem species including beryl. Over the past three decades Pakistan has been providing the world market a steady supply of beryl specimens and gem rough. Pakistan's mineral wealth is explored in Lapis International's monograph *Pakistan: Minerals, Mountains & Majesty* (2004).

Equally as exciting are new finds from well-known localities, such as the fabulous pocket uncovered in the summer of 2005 at **Mount Antero**, Colorado (page 26). New England's old pegmatites (page 28), often forgotten, are also good for a surprise. In 1997, Gary and Mary Freeman began to mine the **Orchard pit**, a then insignificant pegmatite adjacent to the famous **Bennett quarry** in Oxford County near Buckfield, Maine. In 1998, they opened a football-sized pocket containing etched golden beryl. The subsequent years have seen pockets yielding remarkable crystals of gemmy blue and green beryl as well as North America's finest golden beryl!

There is a great deal to share about this and many other of the world's beryl localities and about the mineral beryl and its cousins. Over the next ten or so pages, we will introduce the rarer species in the beryl group beginning with indialite (below), followed by pezzottaite (facing page), bazzite (page 16) and stoppaniite (page 20). Much of the balance of this edition will showcase the spectacular color varieties of beryl and some of the localities that produce them.

Indialite at a Glance *compiled by William "Skip" Simmons & Alexander Falster*

Indialite is dimorphous with cordierite and is extremely rare. Indialite was named for India, the country in which it was first found (Meagher & Gibbs, 1977).

Composition

$Mg_2Al_4Si_5O_{18}$

Dana number: 61.1.1.3; Strunz: 9.CJ.05

Crystal System

Indialite is dimorphous with cordierite. The point group is 6/m 2/m 2/m. The Space group is P6/mcc. The a-dimension is 9.800 Å and the c-dimension is 9.345 Å (Meagher & Gibbs, 1977). The structure contains two tetrahedral sites: T1 which is larger (T1 contains 0.72 Al, 0.28 Si) and the smaller T2 site (which contains 0.30 Al and 0.70 Si).

Optical Properties

Indialite is uniaxial negative. N_O = 1.537-1.539, N_E = 1.532-1.535.

Physical Properties

Indialite has vitreous luster, a Mohs hardness of 7 to $7\frac{1}{2}$. Cleavage is not observed. Density is 2.512 g/cm^3 (measured and calculated).

Color

Indialite is transparent and colorless.

X-ray Diffraction Lines

The following are the eight most intense diffraction lines for indialite (JCPDS # 13-293): 8.48 (100), 4.89 (30), 4.094 (50), 3.379 (55), 3.138 (65), 3.027 (85), 2.640 (25), 1.688 (30).

Forms and Morphology

Indialite occurs as hexagonal grains and sometimes in spiral aggregates and as crystals with multiple terminations.

Alteration and Replacement

No alteration or replacement has been noted.

Locations

The type location for indialite is in the Bokaro coal seam, southwest of Azaribagh, Bihar, India (Meagher & Gibbs, 1977). A second location for indialite exists near Unazuki in Japan (Anthony et al. 1995).

Associated Minerals

Indialite is found in association with enstatite, anorthite, corundum, magnetite, glass (for the Indian location) and cordierite, sillimanite, andalusite, quartz, and biotite at the Japanese location at Unazuki.

Pezzottaite

Federico Pezzotta, curator of the Natural History Museum in Milan, Italy, has been studying Malagasy pegmatites for more than a decade. He recounts the discovery of the new beryl group species that bears his name.

The type locality for pezzottaite, **Ambatovita**, is south of the village of Mandrosonoro, in an undeveloped area of central west Madagascar. I first heard rumors of the place in 1995, when I was a partner in an Italian-Malagasy mining company. At that time, local people spoke of beautiful gemstone tourmaline that came from pegmatites south of Mandosonoro. Intrigued, I hired a Malagasy geologist to assess the potential of the pegmatites. The geologist came back alive (his guide was not so lucky), and though the potential of the gem pegmatites had not been confirmed, the dangerous reputation of the area had.

For a while, nothing more of interest seemed to come out of the region until November 2002, when people from around Mandrosonoro began bringing fragments of large polychrome liddicoatite crystals accompanied by gemmy, partially corroded, pink to green-blue spodumene crystals to Antananarivo in significant quantities. A group of local miners had discovered an extraordinary miarolitic cavity in a pegmatite known as **Sakavalana**, not far from the village of Ambatovita.

At the time, I was in Madagascar to study some pegmatites in the Betafo region. Observing the exceptional tourmaline rough arriving from around Mandrosonoro, French dealer Laurent Thomas (LT) and I resolved to learn more about the locality.

Shortly after returning to Italy, LT received a stock of stones containing a parcel of some tens of very unusual tabular, reddish, hexagonal crystals, 1 to 6 centimeters across. The general appearance of the crystals was somehow similar to that of tabular beryl, but the color was reminiscent of red tourmaline; moreover, the mineral was strongly pleochroic and had a rather high density as well as high refractive indices.

LT sent me an e-mail containing photos and descriptions of the samples. He also sent fragments for analyses. In the meantime, I arranged for a car and guide to take LT to the locality. The trip was planned for the second week of December. In spite of the fact that it was the beginning of the rainy season, LT succeeded in reaching Sakavalana, but Laurent was greeted by gunshots and hastily departed.

In collaboration with my colleagues at the

Pezzottaite with tourmaline on muscovite from the Sakavalana pegmatite near Ambatovita, Madagascar, the overall specimen is 6 cm wide.

Collection and photo Kevin Dixon

Pezzottaite *from the type locality the Sakavalana pegmatite near Ambatovita, Madagascar*

The overall specimen is 4.5 cm wide.

Collection and photo Kevin Dixon

Pezzottaite at a Glance *by William "Skip" Simmons & Alexander Falster*

Pezzottaite is a rare cesium-rich member of the beryl group. Crystals are generally small, reaching a maximum dimension of several centimeters. The color, hardness and the chatoyance of some material have quickly made pezzottaite a highly sought gem material (Simmons et al. 2003; Laurs et al. 2003; Hawthorne et al. 2004). Pezzottaite is named for Dr. Federico Pezzotta (b. 1965) of the Museo Civico, (Milan, Italy) in recognition of his many research contributions on pegmatites and the mineralogy of deposits in both Madagascar and Elba (Hawthorne et al. 2004).

Composition: $Cs(Be_2Li)Al_2Si_6O_{18}$

Crystal System and Structure:
Pezzottaite is rhombohedral; space group R3c; z = 18; a = 15.946 Å; b = 15.946 Å; c = 27.803 Å. The basic structure is the same as beryl with cesium filling the channels and lithium substituting for beryllium (see page 14-15).

Physical Properties:
Pezzottaite has vitreous luster and a Mohs hardness of 8. It is inert to short and longwave ultraviolet radiation. The cleavage on {0001} is imperfect; parting is not observed. It is brittle and shows conchoidal to irregular fractures. Many crystals are penetrated with liquid filled tubular inclusions, which give some stones a pronounced cat's eye effect.

Color:
Pezzottaite is typically deep purplish pink and ranges from transparent to translucent. Dichroism is moderately strong: pinkish orange to pink-orange (ω ray) and purplish pink to pink-purple (ε ray).

Optical Properties:
Pezzottaite is uniaxial negative. N_0 = 1.615-1.619; N_E = 1.607-1.610, birefringence 0.008-0.009.

X-ray Diffraction Lines:
The following are the eight most intense diffraction lines for pezzottaite: 3.271 (100), 3.027 (41), 3.019 (29), 2.871 (52), 2.229 (12), 2.215 (14), 1.749 (12), 1.632 (14) (Simmons et al. 2003; Laurs et al.2003; Hawthorne et al. 2004).

Forms and Morphology:
The crystals are typically tabular to thick tabular.

Alteration and Replacement:
No alteration or replacement has been noted but corrosion on prism faces is fairly common. Corrosion may be followed by an overgrowth of tan-colored pezzottaite with very high cesium content that is filled with numerous fluid inclusions (Laurs et al. 2003; Hawthorne et al. 2004).

Occurrence:
Pezzottaite occurs only in cesium-rich pegmatite environments. At the Sakavalana pegmatite near Ambatovita, Madagascar, it appears to have formed after pollucite was corroded and released a flood of cesium into the late stage pegmatite environment (Simmons et al. 2003; Laurs et al. 2003; Hawthorne et al. 2004).

Locations:
The type location is the Sakavalana pegmatite near Ambatovita, 140 km southwest of Antsirabe, Madagascar (Laurs et al. 2003). A second location in the Deva mine, Paroon Valley, Kunar, Nuristan, Afghanistan has now been confirmed (Laurs et al. 2003).

Associated minerals:
Pezzottaite is found in association with microcline (amazonite), quartz, albite, tourmaline (schorl to elbaite and liddicoatite), spodumene, lithian muscovite, chiavennite, danburite, spessartine, zircon, pyrochlore, cassiterite, hambergite, Mn-hydroxides, and chalcedony (Simmons et al. 2003; Laurs et al. 2003; Hawthorne et al. 2004;

Natural History Museum and at the University of Milan, I had begun to analyze the reddish mineral. Our results evidenced a structure similar to that of beryl, but with a composition enriched with extraordinary levels of cesium, enough cesium to define a new mineral species.

Rumors that this mineral was something valuable were rapidly circulated in Madagascar. By January 2003, prices in the Antsirabe and Antananarivo markets had increased sharply. A few European and US gem dealers were buying the material in quantity, in some cases thinking it was "unusual tourmaline" and in others "raspberry beryl." In spite of the fact that no one really knew what it was, dealers were offering this new stone at Tucson in 2003.

Following its introduction at the Tucson show, gemologists and mineralogists began studying fragments of the mineral. At least two groups of researchers, one from the US and Canada and another from Switzerland, submitted proposals for a new mineral species to the Commission of New Minerals and New Mineral Names of the International Mineralogical Association.

I was both surprised and happy when Skip Simmons, speaking for the US/Canadian team, phoned to ask whether they could suggest to the IMA that the new mineral be named for me.

News of the "fabulous" prices being commanded for the material in Tucson had reached Madagascar, inciting a virtual war among the Malagasy mineral and gemstone dealers, all of whom wanted to handle this new mineral. The fighting abated as the material was sold and out of the country and as it became clear that the pegmatite at Ambatovita had been exhausted.

I returned to Madagascar in March-April 2003, and with the financial backing of some European mineral-collectors, I was able to collect a number of matrix specimens. These specimens enabled us to better characterize the paragenesis, color distribution and the growth and corrosion processes of this new mineral. In early July 2003 with the help of Malagasy guides who negotiated with the locals, *Gems & Gemology* editor Brendan Laurs and I embarked on long and adventurous tour of Madagascar. It was on that trip that I first visited the Sakavalana pegmatite. I also had the opportunity to document the surface and the depth of the pegmatite.

One of the best days of my life was that one in early September 2003 when I learned that the IMA, having received a collaborative proposal by the US/Canadian and Swiss research teams, had approved the new mineral with the name "pezzottaite."

Pezzottaite

Upper: *A 3x3 cm matrix specimen*

Middle Left*: This 1.8 cm wide single crystal exhibits the mineral's corrosion pattern. Both specimens are in the collection of Kevin Dixon, who is also the photographer.*

Middle Right*: A 2.15 (left) and a 3.68-carat gemstone cut from Malagasy pezzottaite rough Private collection; Jeff Scovil photo*

Crystal Structures of

William "Skip" Simmons and Alexander Falster of the University of New Orleans in Louisiana illustrate the basic structures of the two beryl group species beryl and pezzottaite.

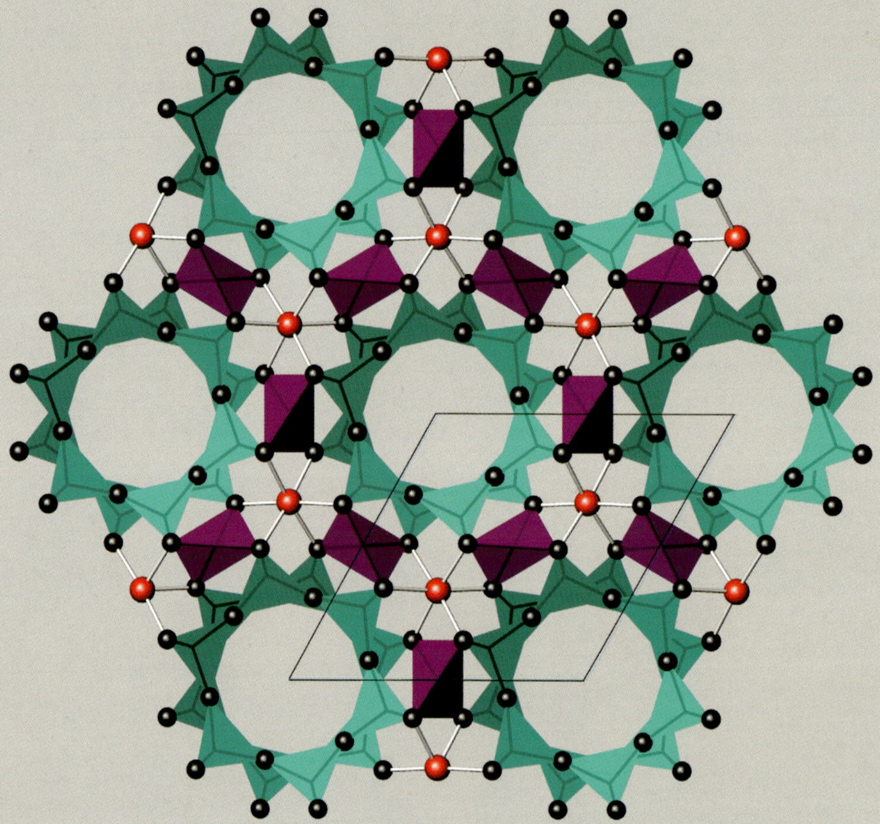

Beryl Structure

*The **above** graphic depicts the beryl structure projected parallel to (0001), including the six-member rings of SiO_4 tetrahedra (blue), the distorted BeO_4 tetrahedra (purple), the octahedrally coordinated aluminum sites (red spheres), oxygen (black spheres), and the large open channels that extend along the* c*-axis. The unit cell is indicated by the black lines. The diagram on the **lower left** shows the beryl structure projected perpendicular to (0001). That in the **lower right** shows the beryl structure projected parallel to (0001) including water molecules (orange spheres) in the channels.*

The diagrams on this page are drawn with the program Atoms *by Eric Dowty using the data of Brown & Mills, (1986).*

Beryl is a cyclosilicate consisting of a three dimensional network of Si_6O_{18} rings. The distinctive aspect of the structure is produced by six-membered rings of SiO_4 tetrahedra (colored blue in the diagrams) that are stacked on top of one another. Aluminum and beryllium atoms lie between the rings, linking them together. The

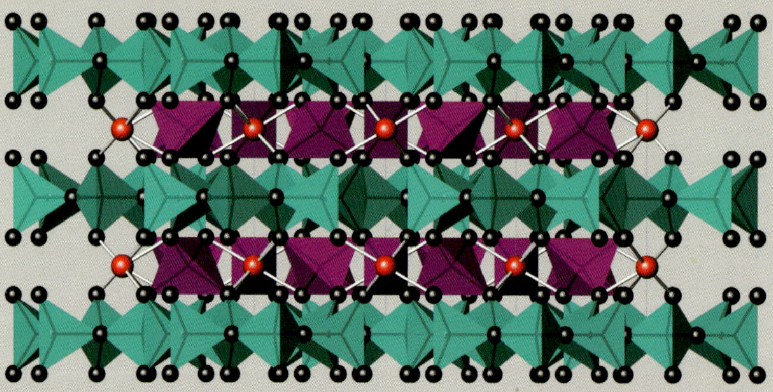

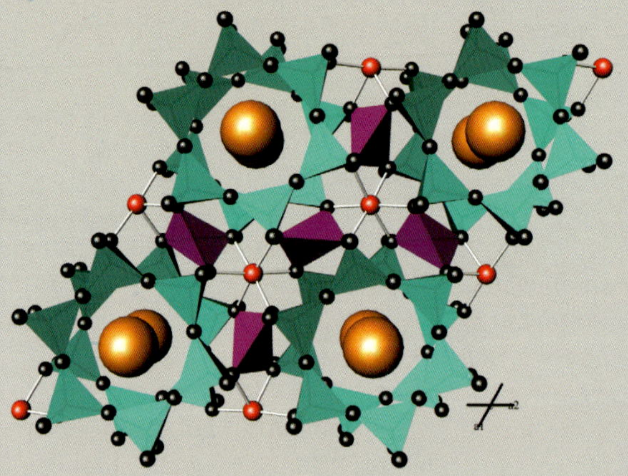

Beryl and Pezzottaite

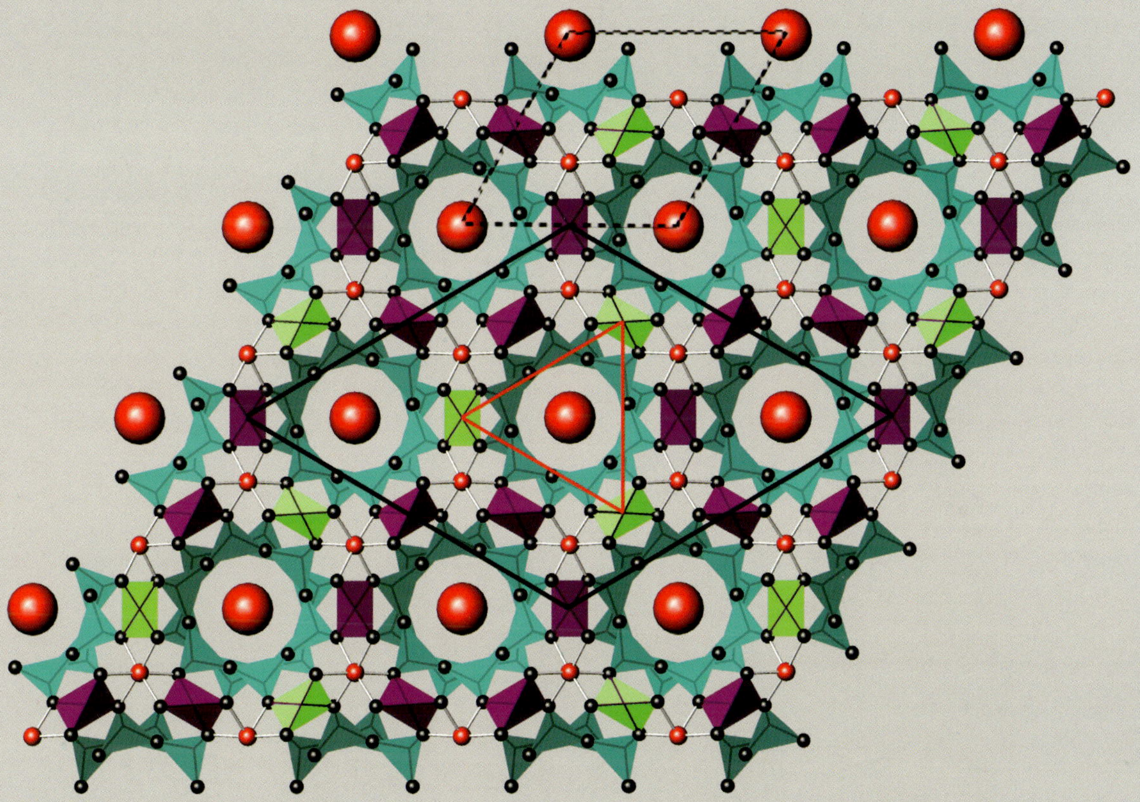

The graphic to the left depicts the crystal structure of pezzottaite projected parallel to (0001). The basic structure is the same as beryl with cesium filling the channels and lithium substituting for beryllium. Cesium atoms are encircled by twelve-membered rings consisting of six BeO_4 tetrahedra and six AlO_6 octahedra. In pezzottaite there are two types of twelve-membered rings. One type is identical to that of beryl and the other type consists of six AlO_6 octahedra, three BeO_4 tetrahedra (purple) and three LiO_4 tetrahedra (green). The dashed lines show the beryl unit cell for comparison to the larger pezzottaite unit cell (solid black lines). The red triangle shows the three-fold symmetry.

This diagram was drawn using the program Atoms *by Eric Dowty.*

aluminum (red) atoms are in octahedral coordination with six surrounding oxygen (black) atoms. These atoms link the rings both laterally and vertically. The six-membered rings of SiO_4 tetrahedra alternate along the *c*-axis with twelve-membered rings consisting of six BeO_4 tetrahedra (purple) and six AlO_6 octahedra to form the crystal structure.

The silicate rings are stacked along the *c*-axis in exact alignment with one another. This arrangement gives rise to continuous channels several Ångstroms wide in the crystal structure where water molecules and other larger alkali ions can fit. The result is like a honeycomb with the open channels running parallel to the *c*-axis.

Water and cations, particularly alkalis such as sodium, lithium, potassium, cesium and rubidium, are contained within the channels. The larger alkali atoms, sodium and cesium, can only fit into the structure in the channel openings, Ginzburg (1955). The smaller alkalis such as lithium substitute for beryllium in the crystal structure. Generally, potassium and rubidium are present only in minor quantities. Water molecules occur in the channels in two orientations, depending on their proximity to alkali ions.

Beryl
Point group: 6/m 2/m 2/m
Space group: P6/mcc
a = 9.205-9.274 Å
c = 9.187-9.249 Å

Pezzottaite
Space group R3c
Z = 18
a = 15.946 Å
b = 15.946 Å
c = 27.803 Å

The crystal structure of pezzottaite is similar to that of beryl, as its structure is also based on the six-membered rings of SiO_4 tetrahedra. In pezzottaite, however, the channels are nearly filled with cesium (large red spheres in the pezzottaite structure **above**), and lithium substitutes for beryllium in one third of the BeO_4 tetrahedra (purple).

These structural differences result in a lower rhombohedral symmetry, with a three-fold *c*-axis instead of a six-fold axis as in beryl, and much larger unit cell parameters.

Bazzite from the Alps

Mineralogist Stefan Weiss of Munich, Germany is the editor-in-chief of **Lapis** *magazine and an avid collector of Alpine minerals.*

Alpine collectors covet beryl, as well as its more rare cousin bazzite, in which the aluminum is mostly replaced by scandium. Bazzite crystals containing trivalent iron are remarkable for their sky blue color. At one time, specimen miners and collectors referred to any deep blue beryl from the granites of central Switzerland as "bazzite," but blue is not a diagnostic color for bazzite, which like beryl, can be almost colorless or even yellowish! Microprobe analysis or precision X-ray diffraction is required to distinguish the two species from one another.

Neither beryl nor bazzite are too picky about the environments in which they form. In the Alps, both form under a variety of conditions that range from pegmatitic melts that crystallized at around 650 degrees Celsius to hydrothermal veins that crystallized at about 200 degrees Celsius. Both of these beryllium silicates seem to grow easily in fissures in contact zones between chemically different rock types.

The type locality of the scandium-beryl bazzite is the quarries near Baveno on the Italian side of Lago Maggiore, where in 1915, Alessandro Bazzi discovered it in the granite vugs. Bazzite was described by mineralogist Ettore Artini (Artini 1915). Further occurrences are known from pegmatite vugs in Heftetjern near Tørdal in the Oslo region of Norway; in Kazakhstan; and in Mount Antero, Colorado in the USA (see page 26). As an Alpine cleft mineral, bazzite is generally associated with granitic rocks, especially in the central massif of Switzerland (Rotondo granite and Fibbia granite-gneiss), while there are only a few minor occurrences in the Austrian and Italian Alps.

Bazzite at a Glance *compiled by William "Skip" Simmons & Alexander Falster*

Bazzite is a rare scandium-dominant member of the beryl group. It is named for its discoverer Alessandro E. Bazzi (1892-1929) (Gaines et al. 1997). Bazzite crystals are generally small, typically in the millimeter range, though they sometimes reach a maximum length of several centimeters. Bazzite is morphologically similar to beryl, and in spite of the small crystal size is a highly desirable mineral that is avidly sought by collectors.

Composition: $Be_3Sc_2Si_6O_{18}$
Dana number: 61.1.1.2; Strunz: 9.CJ.05

Crystal System and Structure:
Bazzite is hexagonal; space group P6/mcc, point group 6/m 2/m 2/m; a = 9.521 Å, c = 9.165 Å. The structure is the same as that of beryl, with the octahedral layer strongly compressed parallel to the c-axis. The Me-O distance in bazzite is 2.080 Å, which is larger than the 1.904 Å in beryl. Its larger cell edge is caused by the flattening of the octahedral layer (Armbruster et al. 1995).

Physical Properties:
Bazzite has vitreous luster, a white streak and is not soluble in mineral acids. Intense blue bazzite is commonly electromagnetic (Gaines et al. 1997). Bazzite has an imperfect cleavage on (0001). Its hardness ranges from 6 ½ to 7; density ranges from 2.77-2.85 g/cm^3 (Anthony et al. 1995).

Color:
Bazzite is most commonly light to dark blue; however, one find in the Swiss Alps yielded pale yellow bazzite crystals (personal communication, H. P. Mueller, Steinhausen, Switzerland, 1985).

Optical Properties:
Bazzite is uniaxial negative. N_0 = 1.623-1.628, N_E = 1.622, birefringence ~0.02 (Anthony et al. 1995).

X-ray Diffraction Lines:
The following are the eight most intense diffraction lines for bazzite (JCPDS # 19-162): 8.2 (70), 3.29 (100), 2.94 (100), 2.56 (50), 2.28 (50), 1.795 (50), 1.754 (70), 1.646 (80).

Forms and Morphology:
Crystals are typically prismatic and exhibit the basal pinacoid {0001} and the hexagonal prism {10$\bar{1}$0} (Anthony et al. 1995).

Alteration and Replacement:
No replacement and alteration products have been reported.

Occurrence:
Bazzite occurs in Alpine vein type assemblages, in granitic pegmatites (Tørdal, Norway) (Juve & Bergstøl, 1990) and in miaroles in granite (Baveno, Italy) (Gramaccioli et al. 2000).

Locations:
Baveno, Italy, in miaroles in granite; from pegmatite sources near Heftetjern, Tørdal, Norway (Juve & Bergstøl, 1990); from La Fibbia, Monte Prosa, and elsewhere in Switzerland; Sugarloaf Mtn., Bethlehem, New Hampshire; from pegmatitic segregations at Mount Antero, Colorado (Anthony et al. 1995).

Associated minerals include quartz, potassium feldspar, albite, muscovite and zeolites for the Baveno location and quartz, albite, chlorite, and others in Alpine veins; the Norwegian pegmatite location includes microcline, quartz, albite, beryl, bertrandite and bavenite.

Upper Right: *A 1 mm high, deep blue bazzite prism with hematite rosettes from Fibbia, Gotthard Pass, Tessin, Switzerland*
Carlo Peterposten collection; Thomas Schüpbach photo

Lower Right: *Bright blue bazzite prisms up to 4 mm with colorless stilbite sheaves on muscovite from the Furka Base Tunnel (West), Wallis, Switzerland*
Bern Natural History Museum collection; Walter Gabriel photo

Below: *A 3 mm long, azure, short prismatic, terminated bazzite, with adularia and quartz from Val Strem, Graubünden, Switzerland.*
Bern Natural History Museum; Thomas Schuepbach photo

Austria

In 1977, the **Gjaidtroghöhe** in the **Grossen Fleisstal** (Big Fleiss Valley) in the state of Carinthia yielded Austria's first and only bazzites as aggregates up to 3 millimeters long of pastel blue prisms in association with anatase, hematite and bavenite (Schebasta et al. 1978).

It is worth mentioning that the state of Carinthia also boasts several beryl localities that are open to collectors. A number of Alpine-type vein beryl localities are known from the **Reisseckgruppe** between **Mölltal** and **Maltatal** that can be easily reached by way of Kolbnitz (Kandutsch, 1998). Not far from the hiking trail to **Riekentörl** (between Big Lake Mühldorf and Little Lake Mühldorf), there are outcrops of aplitic zones of coarsely layered gneiss. These carry narrow smoky quartz-bearing veins that are rich in muscovite and limonitized iron carbonates. One of these veins is not far from the Adlerhorst (Eagle's Nest) hut at the southeastern foot of the Riedbock. These deposits yield tangled masses of light blue beryl needles, occasionally even intense blue prisms that can exceed lengths of 2 centimeters (Kandutsch, 1998).

A belt of similar localities occurs between the **Upper** and **Lower Mooshütte**, over the northwestern flank of the Riedbock and eastwards toward Riekentörl. In the high valley of the "**Goasele**," roughly 1 kilometer south of Lake Mühldorf (big or little Lake Mühldorf?), there is a stretch of layered gneisses with "rock crystal" fissures yielding greenish blue, centimeter-sized beryl prisms, which are often covered by limonite and are not easy to recognize until the rain exposes them on the scree (Kandutsch, 1998).

Switzerland

In his thesis, Henry A. Hänni (Hänni, 1980) presents the most comprehensive information to date on Swiss Alpine-type vein beryl and bazzite.

Tavetsch in the Swiss canton of **Graubünden** is a classic locality for bazzite, especially the area around **Val Strem** (**Stremhörner** or **Piz da Strem**) in the southern Aare granite. Here, bazzite forms short to long prismatic crystals and even some skeletal crystals with hollow cavities (Hänni, 1978). Collector Lucas Monn reports having found 1 to 4 millimeter long bazzite crystals with adularia on an aplite matrix near Val Strem. The area also produces beryl; in fact, dark blue, iron-bearing beryl from Val Strem has been erroneously marketed as "bazzite."

The canton of **Uri** yielded the first find of bazzite—crystals up to 5 millimeters long—in the Swiss Alps. The discovery was made in 1908 by legendary strahler Josef Tresch from the town of Urn. Tresch found the crystals in a small quartz vein in granite-aplite of the **Witenalpkehle** in the **Hinteren Etzli Valley**, but the material was not identified as bazzite until 1939.

Fibbia in the **Gotthard** region in the canton of Tessin is a long known locality not just for hematite "iron roses," adularia, apatite and xenotime, but also for deep blue bazzite. These crystals are generally very small—usually not more than 0.5 to 1 millimeters. Bazzite is typically associated with hematite, among which the tiny dark blue crystals tend to get lost. Exceptionally good material, deep blue prisms to 6 millimeters long, were dug in the 1970s and 80s by strahler Carlo Peterposten of Airolo and collector Luigi Crippa of Giubiasco.

In September 1992, on the **northeast flank of the Fibbia**, in the uppermost Tremola Valley, not far from the Gotthard Highway (Panoramica) and only 0.5 kilometers south of the Gotthard Hospice, Tessin collector Remo Guerra of Biasca discovered a fist-sized vug in an aplite vein in granite-gneiss, right next to a busy hiking trail at an elevation of 2,100 meters! The vug yielded smoky quartz crystals to 7 centimeters, lilac-colored apatite floaters to 2.5 centimeters, hematite "iron roses" to 1 centimeter and two specimens of sky-blue, 3 to 5 millimeter thick, 8 to 11 millimeter long beryl crystals. Though the locality is in the middle of the "bazzite country", X-ray diffraction and EDX analyses have shown this beryl to be scandium-free.

True bazzite "giant crystals" are only known from a single find. In 1982, strahler Bruno da Mocogno from Silen uncovered at **Monte Prosa** what are probably the finest bazzite specimens from the Alps. He found them in a fissure along with smoky quartz, adularia, hematite and xenotime. The specimens consisted of sky blue, doubly terminated, prismatic crystals to 1.5 centimeters (!) long, showing only the basal {001} faces and sitting on altered granite matrix.

The western part of the **Furka Base Tunnel** in Wallis produced one of the Alps' finest bazzite finds: a fist-sized specimen, now housed in the Bern Museum of Natural History, is completely covered by colorless, transparent sheaves of stilbite, muscovite and over one hundred bright blue bazzite prisms up to 7 millimeters long.

Sporadic finds of bazzite, with crystals up to 3.5 mm, some of them cavernous, come from the **Grossen Sidelhorn**, the **Kleinen Furkahorn**, as well as from the **Mittal Road Tunnel** / **Lötschental** (Ruegg et al. 2002).

__Left:__ Bright blue bazzite crystals with colorless stilbite sheaves on muscovite; prisms to 4 mm; Furka Base Tunnel (West), Wallis, Switzerland

Bern Natural History Museum collection; Walter Gabriel photo

__Right:__ Blue prismatic beryl up to 11 mm long with {001} terminations from an aplite fissure in the Rieckentörl area

Helmut Prasnik collection; Stefan Weiss photo

Tips for Collectors

In Alpine clefts, acicular beryl and bazzite are often disguised by a thin red-brown "rust" layer or is coated with a brown-black film of organic humus compounds. At first glance, such crystals can easily be confused with needles of rutile.

Since beryl and bazzite are resistant to acids and alkalis, the organic-stained specimens can be cleaned with ammonia-based glass cleaners or dilute caustic soda. Limonite-stained samples may be cleaned with sodium dithionite or, in extreme cases, with 10 percent hydrochloric acid, which does not attack muscovite. Oxalic acid also dissolves limonite stains, but is not recommended when carbonates or apatite are present in the paragenesis because insoluble calcium oxalate is precipitated, the removal of which requires strong mechanical means such as a high-pressure water gun. Vigorous brushing is not normally advisable, because acicular and thin prismatic beryl/bazzite crystals are quite brittle.

Discovery Stoppaniite

Italian dealer and collector Fabio Tamagnini and two friends were collecting in the field when they discovered the rare new mineral stoppaniite.

December 23, 1994 found Paolo Rossi, Dario Bonetti and I looking for minerals in a volcanic formation known as Ignimbrite C (an ignimbrite is a pyroclastic flow). We were in a quarry near the small town of Capranica in Viterbo Province in Italy's Latium region.

Ignimbrite C sometimes contains a lot of volcanic ejecta, including sanidinites. Sanidinites are ejecta that have hosted the majority of the rare minerals that have recently been discovered in Latium. These rocks contain tiny (rarely more than 1 cm in diameter) mineral-bearing pockets. Because the size of crystals rarely exceeds 1 millimeter, it is generally impossible to identify the minerals contained in these pockets while in the field. Collectors take the sanidinites home and examine them under a binocular microscope.

Sanidinite
A 6 cm wide sanidinitic volcanic ejecta similar to that which hosted the type specimen of stoppaniite.
Collection and photo Edgardo Signoretti

On the day of discovery, we found a few sanidinitic ejecta. One, with a diameter of 35 cm, promised to contain a lot of little pockets. Paolo took a piece of the sanidinite to examine at home, and we put the remainder of the block in the car.

On the following day, Paolo excitedly phoned to tell me that the sample he had taken contained a beautiful microcrystal of "aquamarine." Paolo was particularly happy because at the time he was working to complete his thesis on the minerals of sanidinitic ejecta and to his knowledge beryl had never before been found in them.

In the days that followed, Paolo found quite a few "aquamarine" crystals in the larger portion of the block. Finding enough material for analysis, Paolo reserved some time on the SEM, with the hope that the mineral would prove to be bazzite (the scandium end-member of the beryl group). The SEM analysis resulted in a surprise, however, as Paolo found iron where he had expected scandium. We had found a new mineral that was eventually accepted by the IMA as stoppaniite.

The mineral was named for collector Francesco Stoppani to honor him for the many new minerals and mineral localities that he discovered in Latium.

Stoppaniite at a Glance *compiled by William "Skip" Simmons & Alexander Falster*

Stoppaniite is an exceedingly rare species in the beryl group, having only been found in one location so far. Stoppaniite is the Fe^{3+} dominant analog of beryl. It was named for Dr. Francesco Saverio Stoppani (born 1947) (Della Ventura et al. 2000).

Composition:
$Fe^{3+}_3(Mg,Fe^{2+})Na(Be_6Si_{12})O_{36} \cdot 2\,H_2O$
Dana number: 61.1.1.4; Strunz: 9.CJ.05

Crystal System:
Stoppaniite is hexagonal; P6/mcc;
a = 9.397 Å, c = 9.202 Å

Physical Properties:
Stoppaniite has vitreous luster. Mohs hardness has not been given but is probably 7 to 8. Stoppaniite has a white streak. The mineral is probably brittle and has a {0001} parting. Density is 2.79 g/cm^3 (measured) and 2.85 g/cm^3 (calculated) (Della Ventura et al. 2000).

Color & Luminescence:
Stoppaniite is light blue. Luminescence has not been given, but stoppaniite is likely nonluminescent.

Optical Properties:
Stoppaniite is uniaxial negative.
N_0 = 1.625, N_E = 1.619. Pleochroism is weak (colorless to very light blue).

X-ray Diffraction Lines:
The following are the six most intense diffraction lines for stoppaniite: 8.12 (65), 4.00 (38), 3.278 (100), 2.903 (64), 2.553 (18), 1.752 (18).

Forms and Morphology:
The crystals are typically prismatic and reach 0.5 mm in maximum dimension.

Alteration and Replacement:
No alteration or replacement has been noted.

Occurrence:
Stoppaniite occurs in cavities in volcanic ejecta.

Locations:
Stoppaniite was found in an explosive breccia of the 'C' pyroclastic flow unit at Capranica, Ronciglione, Vico Lake, Viterbo Province, Latium, Italy.

Associated minerals include the typical minerals of the pyroclastic environment.

Paolo Rossi (left) and Fabio Tamagnini, two of the discoverers of stoppaniite, posed together in 2000 at the Lengenbach quarry, Switzerland.
Photo Emanuela Cucchiari

Upper: *View of Ignimbrite C, the pyroclastic flow in which the ejecta that hosted the first and only known stoppaniite was found.*
Photo Edgardo Signoretti

Lower: *One of the few known samples of stoppaniite, this crystal is 0.23 mm high and is from the type locality, Capranica in Viterbo Province, Italy. This sample is in the collection of Paolo Rossi.*
Photo Roberto Pucci

Beryl Color Varieties

The 9.39-carat aquamarine in the ***upper*** *photograph was cut from rough from Mozambique. Collection Mine Design; Photo Jeff Scovil*

The 4.8 cm long terminated crystal of common beryl on quartz, pictured ***lower left****, was found in Miesberg near Saldenburg, Germany. Collection and photo Stefan Weiss*

The 4.2 cm high aquamarine crystal on the ***lower right*** *was found in Filone Silvana, Val Codera, northern Italy. Natural History Museum of Milan; photo Roberto Appiani*

Though poorly defined, varietal names for the mineral species beryl are very commonly used and are deeply embedded in the collector and gemstone literature as well as in the marketplace. In some cases, these terms have been in use for centuries. In most cases, consumers and collectors are more accustomed to stones' being marketed under their varietal names ("aquamarine" or "emerald," for example) than they are to seeing them more precisely labeled simply as "beryl."

Color varieties are poorly defined in part because the small contents of a given chromophore needed to impart vivid color on a stone are not enough to be included in the mineral's chemical formula. While the chromophore does not occur in sufficient quantities to impact the chemical formula, it can be enough to affect properties such as refractive index, conductivity and crystal habit. The ill-defined nature of varietal names creates a problem for the buyer or collector, as there is no clear, "official" distinction between heliodor and green beryl or between green beryl and emerald.

The determination is in fact a judgment call, one that we do not feel is ours to make. We aspire to be a practical reference and have thus allowed the use of the varietal terms in this volume, as buyers will surely see them used in the marketplace. We have not defined the varieties for the authors. Readers will therefore see the same terms used, as they are in the real world, to describe beryls with very different characteristics (i.e., for some authors aquamarine is gem-quality deep blue beryl, for others it is any even faintly blue beryl of any quality). Individual dealers and collectors have legitimately different ideas about the factors that determine the variety of a given stone. For some, paragenesis and habit are as important as (and are less subjective than) color in determining the beryl variety.

As a point of reference, William Skip Simmons and Alexander Falster bravely categorize here the modern color varieties of beryl. —GS/GN

Common beryl is opaque to translucent and generally ranges in color from pale yellow to green to blue. It may form crystals several meters in length. Common beryl is one of the ores of beryllium and is found at many locations worldwide.

Red beryl, also formally known as bixbite, has a remarkably intense raspberry-red color; **emerald**, the most valuable of the beryl varieties, is the deep grass green variety. Both red beryl and emerald are further discussed pages 98 to 99.

Aquamarine, strictly speaking, is the blue, gemmy variety of beryl. The term, however, is used almost as a catchall for every transparent beryl, ranging in color from nearly colorless through the pale greens and into the most intense blues. Deep blue, opaque or translucent beryl is also sometimes referred to as aquamarine.

Natural aquamarine tends to form long, prismatic crystals, but tabular blue crystals are also found. Aquamarine derives its color from divalent or trivalent iron, which controls hue and intensity. It can be heat treated to eliminate green and enhance blue tones. Irradiation typically turns aquamarine yellow.

The name stems from "aqua" (water) and "marinus" (ocean), an allusion to the ocean's blue color. Aquamarine is almost exclusively mined from granitic pegmatites, notably from miarolitic types of the LCT and NYF types of pegmatites (Černý, 1992). Aquamarine typically occurs in the less geochemically evolved subtypes of these pegmatites.

Heliodor or **golden beryl** is the golden to deep yellow variety of gem beryl. Heliodor tends to form long, prismatic crystals. It is often naturally etched and found detached from any matrix. Heliodor derives its color from the presence of trivalent iron. Much of the golden beryl found on the gemstone and specimen markets today is irradiated or heat-treated aquamarine. The heliodor variety of beryl is discussed in greater detail on pages 58 to 62.

Goshenite from Apo Ali Gun, Skardu district, Northern Areas, Pakistan; 3.9 cm high Collection John Taylor; photo Jeff Scovil

Goshenite is the colorless, gemmy variety of beryl. The variety was named for the locality Goshen, Massachusetts, which produced very light colored beryls (Sinkankas, 1981). The name was initially applied to alkali-rich colorless beryls but has been expanded to include all colorless beryl.

Primarily occurring in pegmatites similar to those that produce aquamarine, goshenite often exhibits the typical prismatic habit of beryl. Some locations, Resplendor in Minas Gerais, Brazil (see page 84), for example, produce tabular to platy goshenite crystals. With very low contents of chromophoric ions, heat-treating or radiation does not generally enhance goshenite.

Morganite is a rare and highly desirable pink to orange color variety of beryl that is usually found in pockets within more highly evolved LCT-type pegmatites. This variety of beryl tends to form in tabular crystals. Morganite typically contains elevated amounts of alkali elements and is generally enriched in cesium relative to common beryl. The pink color of morganite is caused by small amounts of manganese. The name "morganite" was first applied to pink beryl from the Pala (California) district (see page 32) by George Kunz, a gemologist with Tiffany & Company in 1910, in honor of the then prominent American banker and mineral collector (and Tiffany customer) J.P. Morgan. **Rosterite** and **vorobievite** are obsolete terms used in older literature that are equivalent to morganite.

Beryl that is cesium and alkali-rich like morganite, but is white instead of pink, is sometimes referred to as **cesian beryl.** Cesian beryl is commonly opaque to translucent, white and of simple morphology. Unlike morganite, this type of beryl contains little or no manganese to act as a chromophore. The Tanco pegmatite in Manitoba, Canada contains hexagonal, short prismatic cesian beryl. The cesian beryl of the Tanco pegmatite contains the highest total alkali oxide content (up to 11 weight percent) known in beryl (Černý and Simpson, 1977).

Upper: *Morganite with tourmaline inclusions; Urucum mine, Galiléia, Minas Gerais, Brazil; 10 cm wide Collection Michael Scott; van Pelt photo*

Lower: *Cesian beryl from Brack prospect in Glastonbury, Connecticut (USA); 8.7 cm wide Collection Agnes Brack; photo Jeff Scovil*

Best of the Rockies

Tony L. Potucek began collecting when he was 8 years old while growing up in Southern Idaho. He currently lives in Arizona and is an avid field collector and connoisseur of fine specimens.

Whether one's interest focuses on field collecting, gemstones or fine mineral specimens, there is a space for beryl in the cabinet of most collectors. The history, variety of colors, range of sizes, as well as the many associations, localities and price points give beryl wide appeal. Those same factors, however, make it a very difficult mineral to cover in 100 pages.

In choosing which localities to include and which specimens to picture, our goal has been to reflect the breadth of possibilities for a beryl collection. As we have previously published issues devoted solely to Pakistan and Madagascar and the beryl variety of emerald, we have not included much on those subjects in this volume. We begin in North America with our glimpse into the world of historic, new, pricey, cheap, micro, macro, gemmy and common beryl. -GS/GN

The color of the blue beryl from Idaho's Sawtooth Mountains rivals that of almost any locality in the world. The above 7.7 cm high aggregate is in the collection of Cal and Kerith Graeber. Photo Jeff Scovil

Two finds have recently raised the bar for beryl quality in United States. One of those was 1869 carat emerald specimen unearthed in late 2003 at the **Rist mine** near Hiddenite in Alexander County, North Carolina. This specimen is by far the best and largest emerald ever found in North America. It is housed in the ever more impressive collection of the Houston Museum of Natural Science.

In July 2004, Steve Brancato, of Salida, Colorado opened a magnificent pocket on his claim on the east-facing flank of the venerable **Mount Antero**, Chaffee County, Colorado (page 26). The specimens of aquamarine, smoky quartz, muscovite and feldspar contained therein (one of which is pictured right) had collectors dropping their jaws when the Collector's Edge unveiled them in Tucson 2005.

The Antero finds invited comparison to another of North America's celebrated aquamarine localities: the **Sawtooth Batholith** in central Idaho.

The Sawtooth is a Tertiary-age granite pluton, which intruded the older Idaho Batholith, a Cretaceous-age granitic rock complex, and older sedimentary and metamorphic rocks. Extensive glaciation has produced the rugged sawtooth-like features that give the mountain range its name.

Aquamarine was first drawn to the attention of the author in the mid-1970s, while working in the Sawtooth Mountains of central Idaho. Members of the Sierra Club, who were rebuilding trails in the rugged mountain range, discovered blue beryl mineralization in float in several drainages. The material led to other discoveries of translucent to clear aquamarine crystals, mostly as single crystals and occasionally in groups associated with granitic matrix. The mineralization occurs in miarolitic cavities and pegmatites.

Aquamarine is found in both Boise and Custer County and is associated with smoky quartz, albite, and microcline. Less common associated minerals are zinnwaldite, spessartine, and hematite, with rare occurrences of bertrandite. The color is truly blue with no apparent green cast to the crystals. The aquamarine crystallizes in both simple flat terminations and as more complex terminated crystals, the latter making for striking specimens. Luster is usually satin-like, but the best of the material is very lustrous, and the crystal interiors are gemmy. I have observed weathered and obscured pockets in which small trees have taken root, and the beryls have been distributed down-slope from the weathered pocket by tree roots and frost heaving. The crystals were intertwined with roots. Pockets vary; containing isolated single crystals to jumbled masses, which, if taken apart carefully, can occasionally be reconstructed to make for attractive multiple intergrown crystal groups.

The most noteworthy pocket to date, erroneously labeled as "Centerville," was a find made by an Idaho Falls collector and dealer in the mid-1970s. These specimens are reported to have originated from a pegmatite. The blue crystals have very high luster. While usually occurring as single crystals, some crystals were found attached to other aquamarines, and some were associated with fairly lustrous crystals of dark smoky quartz.

Beryl on Feldspar with Smoky Quartz and Muscovite
10 cm high x 6.5 cm wide (reconstructed)
Mount Antero, Chaffee County, Colorado, USA
Collector's Edge specimen; Kevin Dixon photo

Mount Antero: The Peak of

Chevron Texaco geophysicist Mark I. Jacobson of Lafayette, Louisiana has extensively collected, researched, lectured and written on Colorado's Mount Antero.

Aquamarine from Colorado's legendary Mount Antero was discovered by "Nels" Nelson D. Wanemaker and was announced in a letter to George F. Kunz in 1885. Continued mining discovered outstanding crystals of phenakite, bertrandite and topaz. Geologically, these minerals formed in pegmatites, greisen, and hydrothermal veins associated with the upper portions and first crystallized parts of the peraluminous Mount Antero granite. The mineralized areas are exposed on the bare-rock talus slopes of Mount Antero and White Mountain, generally above elevations of 3,600 meters.

These Oligocene age (29.8 million years ago) deposits formed during the late stages of the Mt. Aetna volcanic center, which was part of the Early Tertiary volcanic activity common across the western United States. Most of this volcanic activity and granitic plutonism was anorogenic with an NYF geochemical signature and was related to the spreading of the continental crust in the western United States. In Colorado and New Mexico, the Rio Grande rift is evidence of this extension. Geologically similar pegmatite fields with aquamarine occur in Idaho especially the Sawtooth Mountains (page 24) and in Utah.

Aquamarine occurs in miarolitic cavities containing albite, beryl, bertrandite, fluorite, microcline, muscovite, phenakite and quartz; hydrothermal veins containing quartz, beryl, fluorite, muscovite and phenakite; vugs within quartz-microcline pegmatites; molybdenite-quartz veins; and muscovite-beryl-topaz greisen. In the field, these deposit types form a continuum with occurrences often having similar geologic features.

Beryl generally begins to crystallize slightly later than the start of quartz, microcline and albite crystallization. Depending on the cooling history of each vein, phenakite will form either from the partial dissolution of beryl or directly without beryl having previously crystallized. Bertrandite, the rarest of the beryllium suite, is usually found encrusting all the other minerals. The rarest and most desirable mineral specimens have bertrandite, phenakite and adularia encrusting lightly etched aquamarine. Topaz is quite rare and, except for in the greisen deposits, is not found in association with beryl.

No single collector has ever brought large quantities of Mount Antero aquamarine to the collector market. The pockets are too small, often holding only a dozen crystals and require so much handwork that one person cannot open and collect many pockets in the three-month collecting season. Nonetheless, specimens have been trickling into collections and museums since the late 1800s.

Wanemaker specimens from the late 1880s, often of lightly etched single crystals 3 to 6 centimeters in length, now reside in museums in Denver, New York and Chicago. Aquamarine crystals collected by Ed Over between 1928 and 1949, and with Art Montgomery in 1938 are scattered through museums and private collections. The largest crystal (17.9 by 2.86 centimeters) known is dark blue and glassy but is heavily fractured. Ed Over found it in 1932. Median crystal size is 5 by 1 centimeters.

In 1956, Grady Cardwell built a road from Alpine on Chalk Creek up to the south knob of Mount Antero to access his beryllium mining claims. John King and his associates extended that road south to White Mountain to reach their claims. The California mine, a molybdenite-quartz vein to the west of White Mountain, was reopened at that time. The California mine is rich in pockets containing blue to colorless beryl as well as quartz. Some terminated beryl crystals from the California mine have exceeded 8 centimeters in length.

Although the ventures were commercially unsuccessful, the new roads allowed successive generations of prospectors access to the pegmatite areas. Around 1968, multiple pockets from the underground workings at the California mine produced some of Mount Antero's more notable finds of aquamarine. Andrew Taylor, Taranik and Nicolaysen hand mined the ceiling of the main adit and recovered a tabletop-sized lot of aquamarine, including one terminated 8.9-centimeter crystal with smoky quartz. Facet-grade, unterminated 2 to 5 centimeter long crystals were found north of the California mine in the late 1970s, and in 1987, terminated aquamarine crystals reaching 580 carats (8.9 by 3.2 centimeters) were recovered from the Wind

Colorado Beryl Collecting

***Brilliant**!*
*The 7.2 cm high aquamarine crystal pictured **left** was collected in 1992 by Cliff Nicholson and Curtis Abbott.*
*The 1.7 cm high aquamarine on fluorite pictured on the **right** was found in 1978.*
Both specimens are in the collection of Colorado dealer/collector Dave Bunk.

All three photos were taken by Jeff Scovil.

The 6.5 cm high green beryl was purportedly collected by Ed Over during his tenure on the mountain. It resided in the Denver Museum for a time before being traded to a dealer who sold it to the Byer Museum. Collector Lance Hampel bought the Byer collection in 1988, and the specimen resides in his collection today.

Cave Hole on the Snowpatch by Pete McCrery.

Mount Antero continues to produce aquamarine and other gemstone crystals, which can be obtained either by self-collecting or at Colorado mineral shows. All of the mineralized land is public land within the San Isabel National Forest except for the patented California mine. Active mining claims are present on both Mount Antero and White Mountain. Digging on these requires the permission of the claim holders. Inquire at the Bureau of Land Management office in the Denver metropolitan area for the current claim status and regulations.

The deposits will most certainly remain uneconomic for mining beryllium and for large-scale mining operations that require lots of equipment. For the one and two man operation, removing gemstones and well-crystallized specimens by hand with minimal equipment continues to be a rewarding experience.

New England Gem Beryl

Connecticut native Bruce Jarnot has been a noted collector of Connecticut gems and minerals for more than 30 years. Bruce earned his PhD in biochemistry from Wesleyan University and currently works as a toxicologist for the American Petroleum Institute.

From weathered outcrops and plowed fields, beryl crystals have been found in New England since colonial times. In many areas throughout the region, a walk through the woods even today will reveal numerous exposures of granitic pegmatite, scraped bare as Ice Age glaciers advanced across the region. Beryl is widespread in these pegmatites, which are geographically concentrated within regional districts. Major concentrations describe a wide belt, diagonally stretching southwest from the Paris-Rumford and Topsham districts in southern Maine, through the Grafton and Keene districts of New Hampshire, through Massachusetts to the Middletown district in Connecticut. Many scattered individual pegmatite deposits also appear along the extension of this belt, which continues through western Connecticut into the coastal mid-Atlantic States.

Beginning around 1800, early happenstance discoveries gave way to regular and sometimes rich finds of beryl as farmers began to quarry feldspar for use in abrasives and porcelain and sheet mica for use as transparent isinglass. The *Catalog of American Minerals with Their Localities*, published in 1825 by Samuel Robinson, lists 21 different New England localities for beryl.

Sometime in the late 1800s, the old Columbia Gem mine in Springfield, New Hampshire produced this very fine, 2.2 cm high gem beryl crystal now housed in the collection of Bob Whitmore. Photo Jeff Scovil

During this early period of mining, common beryl was sent to the dumps along with quartz and other pegmatite accessory mineral tailings; observant miners, however, often preserved exceptional crystals and gem material. Many fine mineral specimens and gem crystals of beryl were subsequently donated to universities and museum collections that benefited greatly from the generosity of quarry operators. Alternately, if operations yielded sufficient facetable beryl, the sale of gem-quality material provided supplemental revenue for the owners.

Such was the case with fine heliodor and aquamarine from the Slocum quarry and the Roebling mine in Connecticut. The gem rough was sold to Tiffany & Co. and other jewelers in New York City and subsequently marketed as faceted stones. Fine gem beryl and tourmaline from Maine quarries were similarly faceted and sold locally to tourists as souvenir gemstones. Lore holds that both gem beryl and tourmaline from the Strickland quarry in Portland, Connecticut was sold as Maine gemstones on the same Maine tourist market (Richard Schooner, Gino Vitali; independent pers. comm.). By the early 1900s, many New England feldspar and mica quarries ultimately became unprofitable and ceased operation.

World War II restored the demand for sheet mica, and for the first time, common beryl became a valuable strategic source of beryllium. From 1940 to 1945, strong demand for these two minerals brought renewed activity to old feldspar and mica quarries, prompting the opening of a number of new prospects across New England. At about the same time, quarry owners began to capitalize on the widespread boom in the lapidary arts and in mineral collecting by opening their dumps for fee collecting. Fine beryl specimens and gem crystals saved during this second period of pegmatite quarrying in the northeast found a new market with private mineral collectors and mineral shops.

Virtually all beryl-bearing quarries and prospects in New England have produced some gem-grade beryl, with many obscure outcrops and well-known localities represented by one or two faceted stones in personal and museum collections. Only a few localities, however, have yielded significant amounts of beryl in the gem varieties morganite, heliodor and aquamarine. New Hampshire and Maine rank foremost among the New England states for documented production of beryl; Maine, however, notably dominates the list of gem beryl localities. Massachusetts appears limited in the mineralogical literature with reference to historic gem localities. The distinction of being "noted for gem beryl" is somewhat subjective but may be demonstrated by the faceted gems and specimens preserved in museum and personal collections and by references in contemporary literature including newspapers, periodicals and mineralogical publications as well as natural history and geologic surveys.

Wow!

*The Roebling mine in Upper Merryall, Connecticut produced world-class natural golden beryl. The 2 and 4 carat stones pictured **above** are in the collection of Ed Boucher.*

Photo Jeff Scovil

Morganite

In 1989, the **Bennett quarry** in Buckfield, Maine unequivocally claimed first place among New England morganite localities. Two especially notable crystals came to light during several morganite finds that year. First, a morganite crystal collected in September 1989 was later nicknamed "Peach" for its size and yellowish-pink body color. This lustrous and equant 3-inch prism now resides in the Harvard University collection, along with a large section from the second of these notable morganite crystals, the "Rose of Maine." The Rose was a remarkably sharp 31 x 23 centimeter, deeply saturated orange-pink, gem beryl crystal on a showy matrix of large milky quartz crystals, extracted from a pocket in October 1989. The orange hue of the Rose proved unstable in daylight, and much of the crystal had faded to a stable pink when it was ultimately removed from the matrix and broken for gem rough. Numerous cut gems have since been fashioned from pieces of this 48-pound crystal, including a 184-carat cushion-cut stone in the Maine State Museum. As recently as the summer of 2004, mining at **Mount Mica** in Paris, Maine uncovered pocket morganite crystals that yielded light pink gemstones, 2 to 4 carats in size. Also in Maine, the **Emmons quarry** in Greenwood produced large crystals of morganite in years past, with gems to 100 carats reportedly having been cut.

The **Strickland quarry** in Portland, Connecticut, produced deep pink morganite in the late 1800s through the early 1900s. Later collectors, however, found very little of this gem material in the dumps. Pocket beryl from Strickland was sometimes deeply etched, and a morganite crystal on display in the Wesleyan University collection resembles an irregular mass of pink frost. Numerous gems have been cut from yellowish-pink morganite from the **Swanson mine** in Haddam Neck, Connecticut. The **Gillette quarry**, also in Haddam Neck, is known for large euhedral morganite crystals, often showing well-formed first and second-order prism faces. A nearly spherical and crystallographically complex pink crystal from Gillette, two feet in diameter, is now in the Harvard University collection. Tabular morganite pocket crystals were also found at both the Strickland and Gillette quarries. A transparent rose-colored crystal, preserved at Yale University collection, is over six inches in diameter and nearly two inches tall.

Heliodor

One of the premier New England localities for heliodor is the **Slocum quarry** in East Hampton, Connecticut. Slocum was operated for gem beryl in the early 1900s and again in the 1940s and 1950s. A collection of heliodor crystals from the first period of operation—an inch thick and 2 to 3 inches long, well formed, flawless, and rich pure yellow in color—are preserved in the Harvard University collection. Part of another

Slocum Gold

*East Hampton, Connecticut is home not only to Lapis International, but also to a number of gem beryl localities including the Slocum quarry, which yielded the fine 4.5 cm tall gem, golden beryl crystal pictured **right**.*

Collection Bill Larson; photo Jeff Scovil

flawless, pure yellow crystal, collected during the same period, is displayed in the Wesleyan University collection. The second, more sporadic period of mining again produced fine heliodor, though these gem crystals were distinctly chartreuse or greenish-gold in color and often deeply etched perpendicular to the *c*-axis. Another Connecticut locality operated for both heliodor and aquamarine was the **Roebling mine** in the Upper Merryall section of New Milford. Roebling beryl crystals are typically subhedral and poorly formed, edged with a micaceous 'rind' rather than with distinct crystal faces; consequently, this locality did not produce well-formed or flawless heliodor crystal specimens. Roebling beryl colors are deeply saturated, with heliodor grading from bright golden yellow to golden honey brown. The **Trenton quarry** in Topsham, Maine was known for producing fine gemmy crystals of both aquamarine and heliodor.

*Prospecting the Connecticut woods was a passion for the late Walter Behnke (dealer Russ Behnke's father). He is shown **below** in December, 1992 exposing a beryl crystal from a rock overturned at the CCC Camp in Haddam, Connecticut.*

Photo Joel Sweet

Aquamarine

Aquamarine is the most widely reported gem beryl across New England. George Frederick Kunz, the father of American gemology, wrote of aquamarine from the classic New England locality of **Stoneham**, Maine in a series of papers dating from the 1880s. Kunz first noted a partial crystal of aquamarine, badly broken, uncovered in a farmer's field in Stoneham Township. This crystal section, being "5 inches long and 1 inch across, of a fine blue color equal to any from Siberia," was thought likely to yield a 20-carat gem and several smaller stones from 1 to 6 carats if cut. Several miles away in the same township, a large, partially gemmy aquamarine was uncovered in another field. Following these original discoveries in farmers' fields, the **Aldrich quarry** in Stoneham went on to produce a good deal of large and fine gem aquamarine. Within a few years of his first report, Kunz reports that the Stoneham quarry had produced "very fine blue beryls" including "the finest cut beryl ever found in the United States." This famous 134-carat blue-green stone, now in Chicago's Field Museum, plus "over 300 carats of other fine stones" came from one Stoneham aquamarine crystal. Another large 77-carat Stoneham gem aquamarine is preserved in the Harvard University collection.

As mentioned earlier, one famous quarry operated for gem beryl was the **Roebling mine** in the Upper Merryall section of New Milford, Connecticut. While the Roebling mine is primarily known for heliodor, this locality also produced large quantities of aquamarine. The Roebling mine has produced the rough for many faceted blue gemstones over 20 carats in size, with some reportedly approaching 100 carats, of a moderately deep blue to blue-green color, often referred to as "insulator glass" blue.

Old reports note fine aquamarine from the **Columbia Gem mine** in Springfield, New Hampshire. The **Rice mine** in North Groton, New Hampshire produced fractured beryl crystals, which were often gemmy in part. Gem aquamarine is well known from the **Palermo #1 mine** in Rumney, New Hampshire, which has been operated intermittently for mica, beryl and feldspar for over 130 years. In 1973, this property was reopened solely to recover gem beryl and mineral specimens.

Emerald

Several early references to emerald from New England localities refer to stones that do not appear to fit the contemporary definition of emerald. No surviving example of emerald from a northeastern locality, meeting today's definition of a chromium or vanadium-containing, "grass green" beryl, can be found in any museum or private collection. The phrase "meeting today's definition" provides us with a likely answer to this apparent dilemma, since the specificity of today's definition is of fairly recent evolution in comparison to the long history of mining in this region. As late as the 1800s, "emerald" might describe beryl of any color; accounts of "rose-colored emeralds" from **Chesterfield**, Massachusetts confirm that the term "emerald" was then used as a synonym for beryl. In recent years, however, true emerald-hued beryl has been collected from a marble vein at **Old Mine Park** in Trumbull, Connecticut. Samples analyzed in 2003 by Thomas Fehr at the University of Munich, Germany, revealed neither chromium nor vanadium, but trace amounts of nickel, which may be responsible for these schist beryl's green color. Whether these crystals are really emerald is a matter for continued discussion, and it remains unclear whether gem material has been recovered at this locality.

Prospects

The golden era of New England commercial pegmatite mining has passed, and most feldspar and mica mining operations have ceased. Forests have grown to obscure the many well-known prospects and quarries; a golf course on Collins Hill in Portland, Connecticut now offers a great view of the Strickland quarry and its environs. Happily, there are exceptions including the Bennett and Orchard quarries in Buckfield, Maine. The **Orchard quarry** in particular has recently produced spectacular and important finds of pocket beryl, including aquamarine, heliodor, and as a few bicolor (blue/golden) crystals. Remarkably deep blue aquamarine gemstones have been cut quite recently from material from the **Standpipe Hill quarry** in Topsham, Maine. A renewed prospect at the **Tripp mine** in Alstead, New Hampshire has begun to produce deeply hued aquamarine crystals. In the 1990s, **Songo Pond** in Albany, Maine produced aquamarine of the finest deep blue color. Similarly, the working face at the Slocum quarry, East Hampton, Connecticut has been freshly cleared and readied for new production, and who knows what stories will emerge from the field after the next year's collecting season!

Songo Pond

Though the locality has not produced much in the last few years, the Songo Pond aquamarine finds of the 1990s, including the 13.7 cm high crystal pictured ***left****, made a big splash with New England collectors.*

Jan Brownstein collection; Jeff Scovil photo

Beryl from the Pegmatites of Southern California

San Francisco, California collector Jesse Fisher has researched and written extensively on the pegmatites of Southern California.

Over the course of the past century, the complex pegmatites of Southern California have been one of North America's most prolific sources of gem and rare-element minerals. While the region is best known for its production of magnificent specimens of elbaite and as the original source of kunzite (the lilac-colored gem variety of spodumene), Southern California has also yielded much gem and specimen beryl.

The Southern California pegmatite province is located within the mountainous, semi-arid Peninsular Ranges of Southern California. The province contains numerous individual pegmatite districts. It extends from the San Jacinto Mountains in western Riverside County southward through San Diego County and into the northern portion of the Baja peninsula of northwestern Mexico. Fourteen separate pegmatite districts have been recognized in Riverside and San Diego counties. While the total number of individual pegmatites within these districts is in the thousands if not tens of thousands, only a small percentage have ever produced gem or specimen material in any quantity.

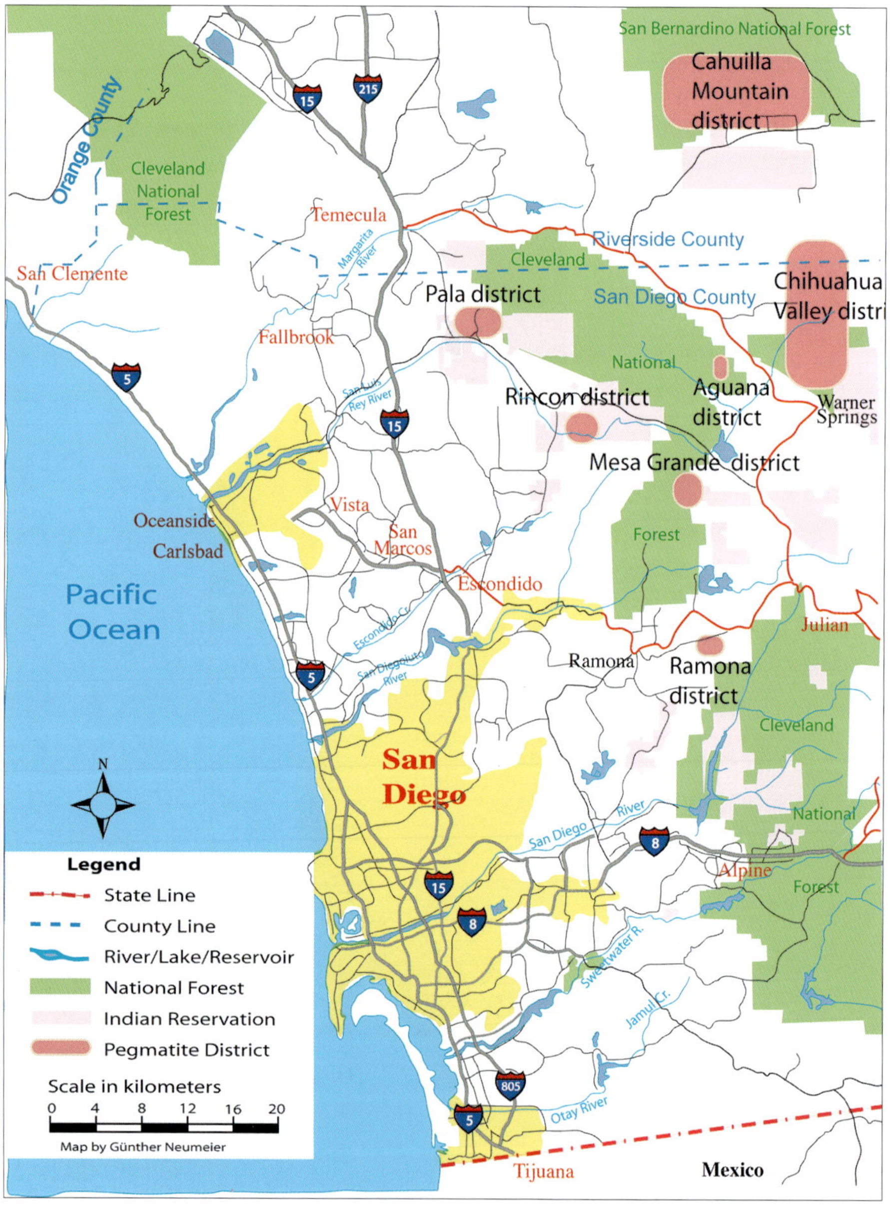

Map of Southern California and Its Pegmatite Districts

The Banner and Jacumba district are east of the region pictured.

The first major find of gem material in Southern California occurred in 1898 at the Himalaya mine in Mesa Grande, California. Immediately after being discovered, the Himalaya pegmatite began producing large quantities of pink and green tourmaline. When word got out about the Himalaya finds, a flurry of prospecting activity began throughout the region. As a result, within the next several years most of the important gem-producing pegmatites in the region had been discovered, and numerous mining claims were established in the Pala, Rincon, Aguanga Mountain, Cahuilla Mountain, Ramona and Jacumba districts.

During the first

decade of the 20th century, Imperial China was a major consumer of California tourmaline due to the popularity of the pink and red varieties with the Dowager Empress Tz'u Hsi. When the Imperial Chinese aristocracy was overthrown in 1911, the demand for California tourmaline in China quickly dried up.

In the U.S., George Kunz and many others had heavily promoted "native" gemstones such as tourmaline, spodumene and beryl, but the loss of the Chinese market and a domestic oversupply of gemstones resulted in the collapse of the local gem market. By 1913 most gem mining operations in Southern California were bankrupt. Gem prices remained depressed for several decades, and between 1913 and the early 1950s, mining activities throughout Southern California were small-scale and sporadic.

By the early 1950s gem prices had begun to rise. This, along with an increased interest in mineral collecting, led to renewed attempts at organized mining, particularly in the Pala, Mesa Grande, and Ramona districts. Mines that have been successfully reopened include the Stewart, Tourmaline Queen, White Queen, Vandenberg, and Katerina in the Pala District; the Himalaya in Mesa Grande; and the Little Three and Hercules in Ramona. In addition, a number of new mines have produced gem and specimen material, particularly in Riverside County and northern Baja California. As a result, the latter half of the 20th century has seen many spectacular finds of Southern California tourmaline, spodumene, beryl, spessartine, and topaz.

A 7.1 cm high blue-green aggregate of beryl crystals from the Fano mine in Riverside County

Al Ordway collection; photo Jeff Scovil

Riverside County Districts

The pegmatite districts of Riverside County make up the northern limits of the Southern California pegmatite province. The districts are poorly defined in comparison to those in San Diego County, and little has been recorded about their mineralogy and history. Much of the area is still remote and sparsely populated, and the exact location of many older mines and prospects is uncertain. Perhaps the best-known pegmatite mine in Riverside County is the **Fano** (or **Fano-Simmons) mine**. First discovered in 1902, the Fano mine consists mostly of surface excavations on a pegmatite dike that lies on top of a ridge located on the northern flanks of Cahuilla Mountain near the town of Anza. Initial attempts at mining produced some pink spodumene and colored tourmaline, but the mine is better known for its more recent but somewhat limited production of well-crystallized specimens of blue gem beryl and schorl. Commercial attempts to work the mine appear to have been limited to the first decade of the 20th century, though hobbyists and high graders later worked the mine. In the summer of 2004, the mine was leased to Ken Gochenour who, along with several partners, has renewed work at the Fano mine.

The **Audry Lynn** is a more recently discovered pegmatite located on the northwest flank of Cahuilla Mountain near Juan Diego Flats. The claim was discovered by Phil Osborne, who worked it for several years in the late 1980s and early 1990s. This prospect produced limited amounts of well-formed aquamarine crystals, both on feldspar matrix and as single crystals; lustrous and often large schorl crystals; and smoky quartz. Specimens of spessartine, cassiterite, and columbite were also recovered. After a hiatus of some years, work has recently resumed at the Audry Lynn and small amounts of gem blue beryl have been found.

This specimen of aquamarine with quartz and albite was recovered in September 2003 by Dana Gochenour and Jim Clanin from the Cryo-genie mine. The beryl crystal is 6.5 cm wide, and the overall specimen is 16 cm across.
Jesse Fisher photo

The Chihuahua Valley District

The **Blue Lady mine** is one of the principal specimen-producing pegmatites in the Chihuahua Valley district. The mine is located in a sparsely inhabited area of north-central San Diego County, just south of the Riverside County border. The district contains numerous pegmatite dikes, none of which have ever been successfully worked on a commercial basis. The Blue Lady pegmatite was discovered in 1905 and was briefly worked for gem tourmaline and cassiterite, but has since been operated solely by mineral collectors. The mine is known largely for its output of well-formed pencil-like crystals of dark blue tourmaline, which occur abundantly in the dike. These tourmaline crystals are rarely if ever of gem quality, however. The mine also occasionally produces specimens of both blue and pink beryl, spessartine, topaz, apatite, and large, well-formed crystals of perthitic microcline. A number of excellent specimens from the mine, including a particularly large and well-formed crystal of peach-colored beryl associated with smoky quartz, are in the collection of the Los Angeles County Museum of Natural History.

The **Cryo-genie mine** is located in rural north-central San Diego County, near the town of Warner Springs. Originally known as the **Lost Valley Truck Trail prospect**, the pegmatite has seen occasional work over the years by hobbyists and amateur collectors. In 2001 the current claim-holder Dana Gochenour, along with his brother Ken, mine engineer Jim Clanin and several others began full-time mining on the property. In the fall of that year, a pocket yielding a number of large, bright pink elbaite specimens along with some blue beryl was encountered. As tunneling continued, a second major tourmaline-bearing pocket was encountered the following year. In September of 2003 a third major pocket was found, this time containing some of the largest and most unique blue gem beryl specimens ever recovered from the pegmatites in the region.

The author took this photo of miners cleaning out a beryl pocket in the Cryo-genie mine in October 2003.

The 4 cm wide morganite crystal with lepidolite, albite, and tourmaline pictured left was formerly in the collection of the late John Barlow.

Collection and photo Jesse Fisher

Beryl from this pocket was unusual in that, while having a medium blue-gray color, the habit was highly tabular, similar to that of morganite and other alkali-rich beryl. Crystals were commonly associated with albite, well-formed quartz crystals, and occasionally pink elbaite. Some of the specimens were quite large, and a crystal dubbed "Big Blue" is one of the most spectacular blue beryl specimens to come from the Southern California pegmatites.

Pala District

The Pala district, located in northwestern San Diego County near the village of Pala, is perhaps the most famous of all the pegmatite districts in Southern California. The district encompasses four hills—Queen Mountain to the west, Chief and Little Chief Mountains in the middle and Hiriart Mountain to the east. These hills are host to hundreds of individual pegmatite dikes. Over the years, numerous mines and prospects have been established on pegmatites in the district, but only a small number of these have ever produced gem or specimen material. The few that have, however, have been the source of some of the world's finest specimens of tourmaline, gem spodumene, and pink beryl. Most of the mines in the Pala district are located within the boundaries of the Pala Indian Reservation. The reservation was established after most of the mining claims in the district, but in recent years many of these claims have become inactive and have reverted to tribal control.

The **Stewart** (or Stewart Lithia) **mine** is located low on the southern slope of Queen Mountain and is one of the earliest pegmatites to have been mined in the region. Early work at the Stewart mine, however, focused on the recovery of lepidolite for its lithium content and not on gemstones. The gem and specimen potential of the pegmatite was not fully realized until it was acquired in 1968 (along with the nearby Tourmaline Queen and Pala Chief mines) by Ed Swoboda, a Los Angeles jeweler and mineral collector. Swoboda, along with partner Bill Larson established Pala Properties International (PPI) to work these mines and market gem rough and mineral specimens. Since that time, the mine has passed through successive operators and to this day continues a low level of production under mine operator Blue Sheppard.

The Stewart mine is best known for its production of vivid "hot" pink tourmaline, but has also been a source of many other well crystallized gem and rare-element minerals, including gem spodumene, pink beryl, columbite-tantalite group minerals, and numerous rare phosphates. Beryl crystals are usually rather pale in color and are sometimes considered goshenite rather than

A tourmaline and beryl-bearing pocket at the Tourmaline Queen mine circa 1973; field of view 1.2 meters
Josie Scripps photo

morganite. Typically, the beryl habit is tabular, with well-developed pinacoid, prism, and pyramid faces. Common associations are with quartz, albite, lepidolite, and colored tourmaline.

Located on the northeast slope near the summit of Queen Mountain, the **Tourmaline Queen mine** is one of the most well known gem-producing pegmatites in California, if not the world. The Queen mine was established in 1903 and until 1912 was, along with the Himalaya mine in the Mesa Grande district, one of the leading sources of gemstones in Southern California. After the collapse of the local gem market, the mine saw little organized work until the late 1960s. Pala Properties International began full time work at the mine during the summer of 1971, and the famous "Blue Cap" pocket was first encountered on December 31 of that year. Tourmaline crystals from this pocket were a rich reddish-pink with a deep blue band at the termination. A number of sizable specimens were recovered in association with quartz and albite, and a few with well-formed, pale crystals of pink beryl. Specimens from this pocket can be seen in many major museum collections in both Europe and North America, and some are among the best-known gem pegmatite specimens ever found. Work by PPI continued at the Queen mine through 1975, and in all at least 10 significant pockets were discovered. The mine has been worked occasionally since then and is still owned by Ed Swoboda.

Though the Queen mine is justly famous for tourmaline, morganite was common in her pockets, sometimes directly associated with tourmaline. Typically, morganite forms tabular crystals with prominent pinacoid faces that are usually glassy. Prism and pyramid faces are generally present and often have a dull or matte surface. The beryl is usually pale pink, and smaller crystals may appear virtually colorless. Pink beryl crystals from the Queen mine are often quite similar in appearance to those from the Stewart mine and distinguishing specimens from the two mines is sometimes difficult.

In 1902 and 1903, the father and son team of Marion and Fred Sickler, along with French Basque prospectors Bernardo Hiriart and Pedro Peiletch, discovered the **Katerina** and most of the other mines on Hiriart Mountain. It is believed that the Katerina mine was the original source of the material identified by George Kunz as the lilac-colored gem variety of spodumene, kunzite, which bears his name (Sinkankas, 1959).

Between 1903 and 1912 the Sicklers, Hiriart and Pelitch worked the Katerina and several other nearby mines from which they reportedly recovered a fair amount of kunzite and beryl (Kunz, 1905; Sinkankas, 1959). Full-time mining came to a halt in 1912, and little was done until the early 1950s when George Ashley began to successfully work both the Katerina and the nearby **Vandenberg mine** for kunzite and morganite.

The Vandenberg was purchased and briefly worked in the early 1970s by Norm Dawson, who produced a modest amount of high quality pink spodumene and beryl. The mining claim on the Vandenberg property was recently ruled invalid by the Bureau of Land Management (BLM), and it has reverted to the Pala Indian tribe.

Serious attempts to work the Katerina began again when Otto Komarek and Byron Weege became partners in the mine in the late 1980s. Work began on the property in 1990, and over the next several years, the pair found a number of pockets including one containing numerous pink beryl specimens, most on a matrix of albite and quartz, as well as a number of kunzite specimens. A dispute with the Pala Indian tribe over the validity of the claim later arose, but in late 1997, the BLM ruled in favor of the mine owners and work on the claim resumed. In November of 1998, a small find of gem pink spodumene was made in the area of the dike worked by Ashley during the 1950s.

The **White Queen mine** is located uphill from

the Katerina mine, just below the summit on the south side of Hiriart Mountain, and has been one of the most important sources of pink beryl in the region. While the White Queen was discovered around 1900, very little mining took place on the property until the late 1950s, when Norm Dawson acquired it.

After driving an initial tunnel of around 8 meters, Dawson encountered a major pocket that produced numerous fine specimens of pink beryl and much gem rough along with many large montmorillonite-included quartz crystals. Several additional pockets of similar composition were found in the mid to late 1960s and again in 1973.

Little subsequent work was done at the White Queen until 1990 when a group of partners including Dawson's sons Bob and Ken began a new tunnel into the pegmatite.

In early September of that year, they encountered a large pocket that produced numerous well-formed specimens of pink beryl associated with albite, quartz, and sometimes muscovite.

Specimens from this pocket are regarded as some of the finest ever produced by the mine and perhaps the entire region. When first removed from the mine, most of the morganite crystals were a peach to salmon pink color. This color, however, changes rapidly to a pure pink on exposure to direct sunlight, and many specimens from the mine have since changed color.

Shortly after this find, a dispute arose with the Pala Indian tribe over the validity of the mining claim. The BLM subsequently ruled the claim invalid, and ownership has reverted to the tribe. No subsequent work has been done at the mine.

The **Elizabeth R mine** works a pegmatite dike located on the north side of Chief Mountain, in the Pala District. The mine is on the site of the former **Ocean View** and **Hazel W prospects**, which were first claimed around 1903. There is no record of production from that time. These claims were purchased and renamed by local miner and collector Roland Reed in 1973. Reed's one-man operation has driven considerable lengths of tunnel.

Specimen recovery from the mine has been highly variable, but in 1982 a major pocket of gem, pink beryl was uncovered. Another find was made in 1992, producing what may be the finest morganite specimen ever found in Southern California. This specimen consisted of a large plate of albite approximately one half meter across, with seven large (up to 15 centimeters across) morganite crystals, and numerous smaller ones scattered across one side of the plate. The beryl crystals were lustrous and well formed and had pale blue cores. This specimen was recently sold by Reed and has since been broken into several smaller specimens. In 2001 Reed sold the Ocean View portion of the property and established a new portal on the Elizabeth R. He has since found a number of modest pockets containing pink beryl, spodumene, and colored tourmaline.

White Queen Beryl

The Houston Museum of Natural History houses this 8 cm high morganite crystal from the White Queen mine.

Jeff Scovil photo

Rincon District

The Rincon pegmatite district is located approximately 17 kilometers southeast of the Pala district in the Pauma Valley of north-central San Diego County. The district encompasses numerous individual pegmatite dikes, most of which are mineralogically simple. A few pegmatites show enrichment in rare elements such as lithium, beryllium, and boron and have produced small amounts of gem and specimen material.

The best-known mines in the Rincon district

are the **Mack**, **Clark** and **Victor mines**, all of which were originally developed by John Mack between 1903-1910. None of these mines proved productive enough to sustain commercial development. And with the exception of some excavations at the Clark mine during WWII for radio-grade quartz, only amateur collectors have since occasionally worked the properties.

The Mack mine is best known for producing small amounts of gemmy, pale blue to blue-green, prismatic beryl crystals. The Clark mine has produced, in addition to quartz, small quantities of gem pink beryl. The Victor mine is best known for producing gem-quality pencil-like crystals of blue to violet tourmaline, but pinks and greens have been found as well, along with a few blue beryl crystals.

Mesa Grande District

The **Himalaya mine** is historically the most productive of the numerous pegmatite mines in the Mesa Grande district. Since its discovery in 1898, activity at the mine has been highly variable, depending largely on the economics of the gem and specimen market. Over the course of the mine's history, however, the production of gem and specimen material has been considerable, and the mine certainly ranked as the largest producer of gem and specimen-grade tourmaline in North America and possibly as one of the most productive gem pegmatites in the world. As with most other gem mines in the region, organized activity at the Himalaya came to a halt in 1912. Mining remained episodic until Ralph Potter acquired the property in the mid-1950s. Potter successfully rehabilitated many of the old workings and found several major tourmaline-bearing pockets in the late 1950s and early 1960s. In the late 1970s the mine was leased and later purchased by Pala International, who operated the mine with considerable success until its closure in 1998.

The main portal of the Elizabeth-R mine as seen in November 2001; Jesse Fisher photo

For all of its mineral richness, however, the Himalaya dike system produces very little beryl. Frozen masses of both pink and pale blue beryl have been observed in the intermediate zones of the main pegmatite, but pocket occurrences in the mine are uncommon. Highly etched fragments of sometimes quite large, pink beryl have occasionally been found in pockets. These fragments almost never show any crystal faces, but are usually quite gemmy and have a rich peach-pink color. Good examples may be seen in the Houston Museum of Natural Science and the Natural History Museum of Los Angeles County.

Secondary beryllium-containing minerals including hambergite and bavenite are occasionally found in pockets in the main dike. These likely formed at the expense of earlier beryl. Hambergite usually occurs as small porcelain-white, wedge-shaped crystals, and is often associated with other pocket minerals such as elbaite and feldspars. Bavenite usually occurs as crusts of acicular white microcrystals, coating earlier formed pocket minerals. Bavenite pseudomorphs after beryl are occasionally found.

Ramona District

The **Little Three** and **Hercules mines** are located in the Ramona district, approximately 3 kilometers east of the town of Ramona in central San Diego County. Both mines were first claimed in 1903 and worked actively until around 1912. They yielded blue and colorless topaz, spessartine, green and pink beryl, dark green tourmaline, and large quartz crystals. The properties were largely idle until the mid-1950s when mining was resumed at the Little Three by Louis Spaulding, Sr. and later by his son Louis Spaulding, Jr. The Little Three mine is best known for producing some of the region's finest natural blue topaz crystals; major pockets were discovered in 1976 and 1986.

The Hercules mine (also called the Hercules-Spessartine mine) is located on an adjacent pegmatite and is best known for producing bright orange spessartine. Though the Little Three and Hercules are separate claims, Louis Spaulding Jr. has worked them both in recent years. During the 1970s and 1980s, Spaulding alternated work between the two dikes.

Beryl is not uncommon in either the main Little Three or the Hercules dikes, although the best specimens appear to have come from the Hercules. Prismatic crystals of pink, blue and colorless beryl up to 8 centimeters long have been found in recent years, the majority being from one 1985 pocket in the Hercules. Beryl crystals are often lustrous and well formed, but highly etched crystals are also found. While crystals of morganite from other occurrences in Southern California usually have a tabular habit, Ramona district crystals tend to be prismatic.

Louis Spaulding Jr. recovered this 2.5 cm tall beryl group from the Hercules dike of the Little 3 mine in 1983.

Collection and photo Jesse Fisher

Jacumba Pegmatite District

The **Pack Rat mine** is the principal prospect in the Jacumba pegmatite district, located in an arid and remote region of southeastern San Diego County, approximately 100 kilometers east of the city of San Diego and just north of the Mexican border. Little mining appears to have been done in the district until late 1974 when local mineral collector Loren Beebe discovered a pocket of spodumene while prospecting the area. A claim known as the "**Beebe Hole**" was established on the site, but little else appears to have been found there.

In 1981 Fred Stevens began working the Pack Rat mine, a claim adjacent to the Beebe Hole on the same pegmatite. Stevens worked the claim for about 10 years, producing some spessartine and apatite crystals, before selling it to Gary Wallace. Wallace has continued to work the Pack Rat on a part-time basis and during the early 1990s acquired the Beebe Hole claim, which he merged with the Pack Rat. The mine has produced an interesting suite of minerals in recent years, including pale blue to green beryl, Alpine-type smoky quartz crystals, spessartine, danburite, columbite, cassiterite, helvite, and numerous phosphates. The pegmatite shows considerable potential for specimen production, and several other mining claims have been established on neighboring pegmatites in recent years. Unfortunately, the Bureau of Land Management has placed severe restrictions on the scale of mining activities that can be conducted in the area, limiting the prospects for future production.

4.5 cm ***(right)*** *and 3.5 cm tall (****left****) beryl crystals from the Pack Rat mine in San Diego County*

Collection and photo Jesse Fisher

Emerald and Aquamarine in Canada

Lee A. Groat of Vancouver, British Columbia is a professor of mineralogy and crystallography at the University of British Columbia. He is also coeditor of the **American Mineralogist.**

The Dog Days of Summer
Fresh snow dusts Tsa da Glisza (Regal Ridge) in this August 1, 2002 view of the locality taken by the author.

Tsa da Glisza (formerly Regal Ridge), Yukon Territory

Late in the summer of 1998 while looking for sulfide mineralization, geologist William Wengzynowski discovered emerald at Regal Ridge, now known as **Tsa da Glisza**, in the southeastern Yukon Territory (map page 43). Bill had a bad knee and was moving more slowly than normal; in addition, he was searching in the talus on the north side of a ridge instead of running the ridge top. When he first saw the green color, he thought it might be malachite but as soon as he saw the first hexagonal cross-section he knew that he had found beryl.

Mineralization at Tsa da Glisza is associated with quartz veins and aplite dikes that intrude metavolcanic rocks. Pale green to green-blue to "emerald green" beryl crystals up to 4 centimeters in length occur in 12 mineralized zones within a 900 by 900 meter area (photo page 42). The beryl's green color is derived predominantly from chromium. Some of the smaller crystals and sections of larger crystals are gem-quality, and a number of gemstones, up to about 2.69 carats, have been fashioned from Tsa da Glisza samples (photo facing page). The property currently belongs to True North Gems Inc., which is publicly traded on the TSX Venture exchange (symbol TGX). The area is snow-covered most of the year, and is accessible in summer via small fixed-wing aircraft and helicopter.

The main host rock for beryl mineralization is a chlorite-plagioclase schist. Geochemical analyses show that the schist is a high-calcium boninite (high-magnesium andesite to basalt); its chromium content (average 960 ppm) makes it the most likely source for the chromium in the emerald.

The occurrence is underlain, at a depth of at least 300 meters, by a two-mica (biotite >> muscovite) granite pluton. Geochemical results (Neufeld et al. 2003) confirm that the pluton is an evolved peraluminous S-type granite. Although the beryllium content of the granite is not anomalously high, the close proximity and consistent correlation of beryllium with tin, tungsten and bismuth in the geochemical data indicate that the granite is the source of the beryllium. Numerous aplitic leuco-granite dikes

Green Eyes

*Cleopatra would have been seeing green had the Tsa da Glisza rough **(left)** and cut stones **(below)** been available on the Ancient emerald market. The largest stone shown below is 2.4 carats.*

Both photos courtesy of True North Gems, Inc.

ranging in size from 40 centimeters to 5 meters in width occur on the property, mostly within the schist.

Quartz veins are abundant throughout the property. Early veins are typically thin, foliation-parallel, sulfide-rich, and contain no tourmaline. All of the other quartz veins are associated with at least some tourmaline, either within the veins or in the vein selvages. When in association with beryl and/or emerald, the tourmaline is dravite to uvite; otherwise, it is schorl.

Emerald is found associated with veins at several different orientations. Mineralization appears to be particularly well developed in the areas of intersection between veins. Emerald occurs most commonly along the margins of quartz veins, as well as within the quartz veins themselves, and in highly altered zones within the schist that extend past the veins and are interpreted to be fluid pathways.

Geochronology confirms that mineralization occurred at the same time as regional deformation and metamorphism related to intrusion of the granite pluton (Neufeld et al. 2004). In particular, the mineralization coincided with the waning stages of granite emplacement. Some of the emerald-bearing veins have been affected by late ductile deformation, as evidenced by the presence of fractures and secondary fluid inclusions in emerald and micro-boudinage of tourmaline grains within vein quartz. In addition, some of the emerald crystals have a white rim, possibly the

result of corrosion by late-stage fluids. At least two of the aplite dikes contain beryl or emerald, which suggests that there is a continuum from the granite intrusion through aplite dikes to beryl-bearing quartz veins (Neufeld et al. 2003, 2004). The aplite and quartz vein geochemistry suggest that beryllium was carried within the mineralizing fluids as fluoride and hydroxide complexes. Chromium entered the fluid system through interaction of the beryllium-bearing fluids either with hydrothermal fluids that had percolated through the host and extracted some chromium, or via element exchange during metasomatic alteration of the vein selvages by the mineralizing fluids. Oxygen and hydrogen isotopic data show evidence of isotopic mixing between the schist and intruding fluids for the beryl-bearing veins;

Above: *A tablespoon provides the scale in sizing up an emerald-bearing "pocket" uncovered at Tsa da Glisza.*

Photo courtesy of True North Gems, Inc.

Right: *The Lened occurrence yielded this 2 cm long green beryl found frozen in quartz.*

Photo by the author

whereas the non-mineralized veins show no evidence for mixing and are likely increasingly hydrothermal in origin (Neufeld et al. 2004).

In August of 2004, the name of the Regal Ridge property was formally changed to Tsa da Glisza (pronounced "sa da-glis-sa"), as the result of a mutual agreement between True North Gems and the Ross River native people. The name Tsa da Glisza was drawn from the Kaska language and means "green stone."

Approximately 33 kilograms of emerald was recovered from Tsa da Glisza between 1998 and 2003. The majority of the emerald recovered is in the form of rough, though some specimens have been found. Most of the material is sent for fashioning into faceted stones, cabochons, beads, and other similar products, though none of the finished products have as yet been sold. Tsa da Glisza is an advanced exploration property and is not in production, though True North hopes to convert the property into a viable mine.

Polished goods produced to date are part of the economic analysis of project feasibility, and neither rough nor polished goods are available for purchase at this time. Likewise, specimens are not offered for sale or donation at this time, though the Pacific Museum of the Earth in Vancouver, BC has occasionally hosted exhibits of the material.

Lened, Northwest Territories

Transparent pale-green beryl crystals may have been discovered at the **Lened** property in the late 1970s, but were rediscovered by Ron Berdahl, the current claim holder, in 1997. The occurrence is located north of the Yukon-Northwest Territories border, 55 kilometers northwest of the ghost town of Tungsten in the Logan Mountains.

Beryl at the Lened emerald occurrence is found within quartz-carbonate veins. The veins (2-30 centimeters wide) extend perpendicularly from a thrust fault across a skarn and pinch out in overlying rhythmically-bedded limestone. The black shale footwall unit is devoid of emerald mineralization. The beryl is intergrown with quartz and calcite adjacent to the wall rock and concentrations are highest near the vein-fault intersections. Beryl crystals from Lened are typically euhedral, up to 3 centimeters in size and 3 millimeters in diameter, although the longest crystals tend to be quite narrow. Most of the crystals are pale green to yellow in color, but a small fraction are a dark grassy-green color.

Chemical analyses indicate that the emerald is richer in vanadium (up to 0.5 weight percent V_2O_3) than in chromium (0.04 weight percent Cr_2O_3). The host skarn ranges from a pyroxene-garnet-amphibole facies at the fault contact with the shale, to a distal pyroxene-garnet facies.

The probable source of beryllium is the Lened pluton. The core contains slightly more beryllium (~8 ppm) than the outer zones (~5 ppm). The probable source of the vanadium is the black

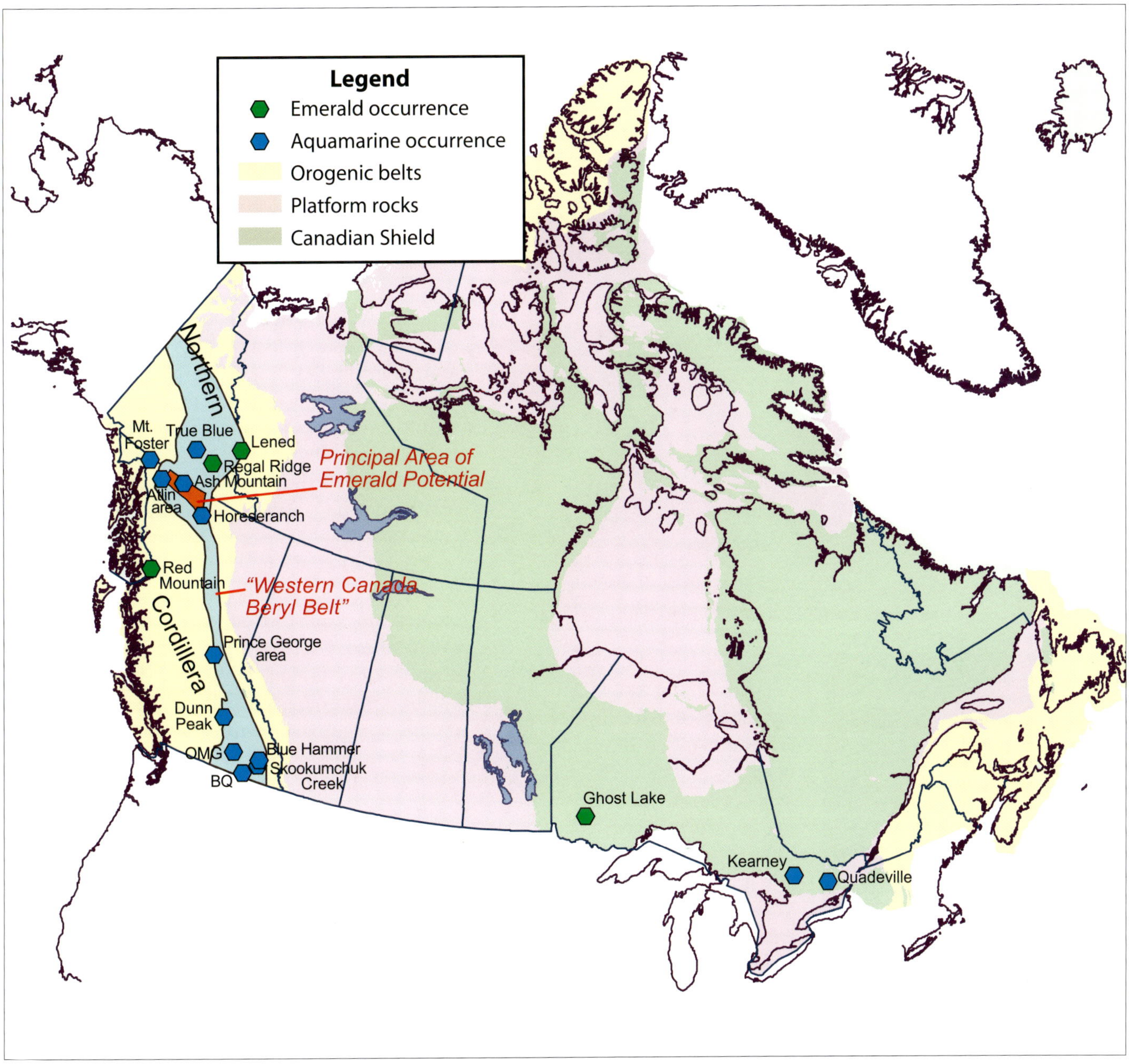

shale, given that it contains concentrations of vanadium greater than 3,000 ppm.

Marshall et al. (2004) suggested that hot fluids from the cooling Lened pluton, traveling along the thrust, produced the contact metamorphic rocks. Prolonged pluton emplacement and cooling resulted in further deformation of the country rocks, with the calc-silicate rocks behaving in brittle fashion and the unmetamorphosed rocks behaving in a ductile fashion. This resulted in a set of sub-parallel fractures in the contact metamorphic rocks. These fractures became conduits for hydrothermal fluids expelled from the cooling granitic rocks; quartz and beryl precipitated in these fractures, resulting in the observed series of sub-parallel veins.

Ontario

The **Taylor 2** beryl occurrence near Dryden in northwestern Ontario was discovered by J.G. Taylor around 1940 (Breaks, 1989). It is presently under option to True North Gems Inc., which refers to it as the Ghost Lake property. At Ghost Lake, beryl occurs within a 3.5 by 1.5 kilometer area of pegmatitic granites (Breaks and Janes, 1991).

The beryl is concentrated in two pegmatite dikes situated 340 meters south of the Ghost Lake batholith. Only the Taylor 2 pegmatite contains emerald (Breaks, 1989). The occurrence is found at the eastern end of a large conformable sill of altered ultramafic rock (peridotite; Satterley 1941) that is exposed as an east-west trending

*The **above** map plots the emerald and aquamarine occurrences described in the text and shows western Canada's beryl belt (after Legun, 2004).*

Ghost Lake Emerald

The Taylor 2 emerald occurrence provided the rough for the 2.34-carat cab ***(pictured right)****, cut, owned and photographed by Brad Wilson. The same locality produced the 2.5 cm long emerald crystal shown* ***below middle****, which is also in the collection of Brad Wilson.*

Middle photo Bernie Gaboury

ridge, approximately 80 meters long and 10 meters wide. The pegmatite consists of three separate limbs that intrude a wide zone of chlorite schist, which sits between rusty metavolcanic biotite-feldspar schist and the end of the ultramafic sill (Garland, 2004). Most of the beryl and emerald occurs within a "zone of mixing" between the southern and central limbs of the pegmatite (Garland, 2004). The "zone of mixing" consists of relict K-feldspar crystals from the pegmatite in a matrix of albite, biotite, and tourmaline. The beryl occurs as euhedral crystals up to 2.3 x 1.8 centimeters in the dark matrix material proximal to the relict feldspar crystals (Garland, 2004). Most of the crystals are opaque white to pale green, but about 10 percent of them are emerald green in color. Electron microprobe analyses show up to 0.47 weight percent Cr_2O_3, 0.05 weight percent V_2O_3, and 0.50 weight percent FeO. Some of the crystals contain inclusions of very fine-grained tourmaline or euhedral, green-blue apatite, and some are veined by fine-grained, equigranular apatite, evidence of albitic replacement (Breaks, 1989).

Still under exploration, Taylor 2 has not produced enough material to allow for a stone count, and with the exception of a few stones cut by Brad Wilson, the bulk of the material recovered to date remains in rough form. A small amount of specimen grade material has been found among the rough. The relatively small quantity of gem-grade beryl recovered from Taylor 2 makes the deposit seem a less than promising commercial gemstone producer.

According to Parsons (1934), large crystals of beryl were found with "amazonstone" and rose quartz near **Quadeville** in Lyndoch Township, Renfrew County. The majority of the beryl was flawed, but a portion of some of the crystals was found to be transparent and free of cracks. In such cases a few greenish-blue aquamarines were cut. Three of these can currently be found in the Canadian National Gem Collection in Ottawa (3.09, 1.80, and 1.35-carat) (Wight, 1986).

Waite (1945) describes an aquamarine from near **Kearney**, Butt Township, Nipissing District weighing 2.1 carats and "flawed but with fine colour;" this is also in the National Gem Collection (Wight, 1986). He further describes a 3.1-carat aquamarine from the Quadeville occurrence (probably the 3.09 carat stone described by Wight, 1986) and states that "the colour and clarity of these stones matches the best from Brazil."

True Blue, Yukon Territory

Also owned by True North Gems, the **True Blue** property is located in the southern Yukon

Left: *Brad Wilson is seen here working the Taylor 2 beryl locality in the summer of 2004. Photo Dawn Kellett*

Territory, southwest of the Tintina Fault. Beryl was discovered on the property in 1976, and a team of geologists working for Archer, Cathro & Associates found dark blue aquamarine there in August 2003 (Rohtert et al. 2003).

Crystals occur in a swarm of closely spaced quartz and/or siderite and/or fluorite and/or tourmaline veins that fill tension gashes in a syenite stock. The veins range in thickness from 0.5 to 20 centimeters and locally comprise up to 30 percent of the rock. The vein zone outcrops in a 700 by 400 meter area and is exposed over an elevation range of 100 meters. Within this area, more than 100 individual occurrences have been discovered. The vein zone is developed near the upper contact between the syenite intrusive body and the pelitic and carbonate country rocks. Syenite is sodic in composition (~8 weight percent Na_2O) and contains moderately high concentrations of beryllium (to ~10 ppm), fluorine (to ~4000 ppm) and rare-earth elements.

Beryl crystals from this property range in size from a few millimeters to 5 by 2.5 centimeters, and in color from pale to medium green and from pale to dark blue. Some of the crystals, especially those occurring with tourmaline, show a blue core and green rim. Because of its rarity, the dark blue material is the focus of efforts in the field. Individual crystals ranged up to 38 millimeters long and 11 millimeters in diameter, and five stones weighing up to 0.82 carats have been cut to

True Blue

Left: *This 4 mm long aquamarine crystal with quartz and fluorite is from the True Blue property in the Canadian Yukon. Photo by the author*

Below: *As of summer 2004 when the author snapped this shot, there was no sign of urban sprawl surrounding the True Blue exploration camp.*

date. According to Rohtert et al. (2003), the stones' color is maintained at exceptionally small sizes for aquamarine. The primary constraints on the size of the stones faceted are the abundant fractures present in the material gathered so far. Microscopic examination revealed that all five samples were also fairly heavily included, which affects their transparency (Rohtert et al. 2003).

Electron microprobe analyses of the dark blue material show high concentrations of iron (to 5.92 weight percent FeO), magnesium (to 3.27 weight percent MgO) and sodium (to 2.51 weight percent Na_2O). The iron concentration is the highest we know of for true beryl. The only member of the beryl family with more iron is stoppaniite; however, stoppaniite is light blue in color, probably because all of the iron is in the ferric state.

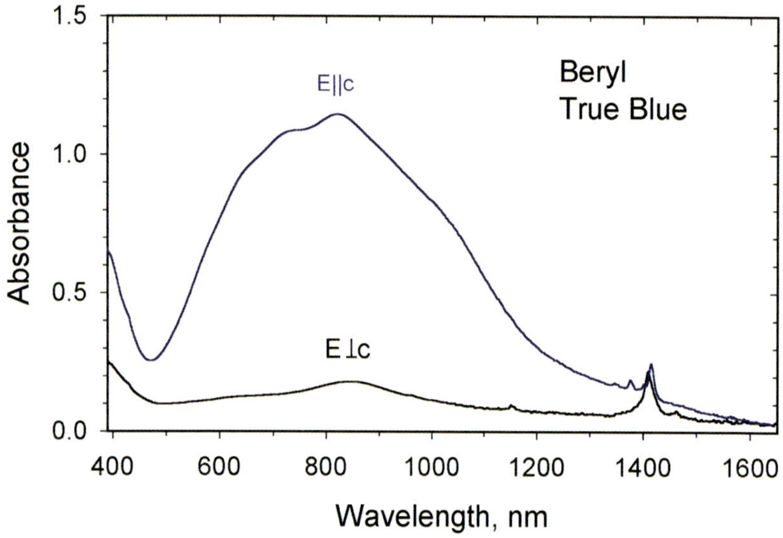

Above: *Absorbance spectra for the True Blue aquamarine, courtesy of George R. Rossman*

Although rare, dark blue aquamarine crystals have also been recovered from the Ambositra beryl-columbite pegmatite in Madagascar (Pezzotta, 2001). Proctor (1984) describes the discovery (in 1964) of eluvial dark blue aquamarine fragments totaling 22 kilograms in the Marambaia Valley in Brazil; four of the crystal fragments were subsequently sold to a dealer for \$45,000. Proctor (1984) also describes a 34.7-kilogram crystal, dubbed the "Marta Rocha," that ultimately yielded 57,200 carats of dark blue aquamarine. Other localities include Lone Pine, California, the Tatu mine in Brazil and Lajes Pintada, Rio Grande do Norte, Brazil. There is some evidence that the True Blue material is darker than the beryl from either California or Brazil. "Maxixe" is the varietal term for dark blue gem beryl from Brazil, the color of which is the result of radiation-induced color centers and fades with prolonged exposure to light (Nassua et al. 1976).

Subsequent to the discovery of the dark blue aquamarine, it became apparent that the mechanism responsible for color in beryl is not completely understood. In the case of the True Blue material, it almost certainly involves intervalence charge-transfer (IVCT) between Fe^{2+} and Fe^{3+} cations. An absorption spectrum of the dark blue aquamarine shows greatest absorption when the electric vector (E) is parallel to the *c*-axis. This suggests that the Fe atoms involved in IVCT are arranged along (or sub-parallel to) the *c*-axis; however, none of the conventional sites in the beryl structure are aligned in this way.

One possibility is that the blue color is due to IVCT between Fe^{2+} cations at the Al sites and small amounts of Fe^{3+} at the 6*i* position, a location with sixfold (trigonal prismatic) coordination that lies between the Al sites and which is normally empty (Platonov et al. 1979); however, there is no evidence of this from X-ray single-crystal data. We are presently using Mössbauer spectroscopy and neutron single-crystal diffraction to study the cause(s) of color in aquamarine from the True Blue property.

British Columbia

According to collector, prospector and gem cutter Brad Wilson (1999), crystals of emerald-green beryl were discovered in 1989 at **Red Mountain**, near Stewart on the central coast of British Columbia. The beryl occurs as small opaque crystals with numerous fractures in narrow quartz-calcite-pyrite veins cutting volcaniclastic rocks adjacent to a quartz-monzonite intrusion. Electron microprobe analyses show this beryl to contain 1612 parts per million vanadium and 1.04 weight percent FeO, with no detectable chromium.

Wilson (1999) also reported on aquamarine from **Mount Foster** near Bennett (photos facing page), the **Horseranch Range** (photo below) near McDame and the **B-Q claims** near Passmore. The aquamarine at Mount Foster occurs in miarolitic

Right: *0.33 ct (left) and 0.21 ct (right) gemstones cut from beryl found in 1993 in the Horseranch Range; collection and photo Brad Wilson*

cavities with microcline, phenakite (Be_2SiO_4), smoky quartz, topaz and gem-quality fluorite. The aquamarine is colorless to light blue and at the time of this writing the two largest stones cut weighed 8.63 and 2.99 carats. The larger stone had many inclusions, but smaller stones weighing less than 2 carats were described as virtually flawless (Wilson, 1999).

Beryl occurs in the **Horseranch Range** as euhedral crystals embedded in granite pegmatite dikes that crosscut high-grade metamorphic gneiss. Although most of the beryl is fractured and opaque, a few transparent crystals have been found. Cut stones are colorless to very pale blue; the largest at the time of this writing weighed about 1 carat (Wilson, 1999).

At the **B-Q** claims, beryl occurs with dark red garnet, smoky quartz, and black tourmaline in miarolitic cavities in a granite pegmatite dike that cuts across the foliation of a high-grade metamorphic gneiss (Wilson, 1999). According to Wilson (1999), the largest aquamarine crystal discovered was 3.3 by 3.2 by 2.3 centimeters with the largest faceted stone weighing 5.26 carats, but Brown (2004) reports that gem-quality crystals up to 10 centimeters in length have been recovered from the pegmatite. Cut gems are pale blue-green in color and tend to have numerous planes of inclusions oriented parallel to the *c*-axis.

Wilson (1999) also reported that the following locations have produced aquamarine crystals that either have been cut or could be cut into gems under 0.3 carats: **Ash Mountain** near McDame, **Dunn Peak** near Barriere, **Skookumchuck Creek** near Kimberley and in the Atlin and Prince George areas.

Two aquamarine crystals more than 8 millimeters in diameter and dozens of transparent to translucent crystals less than 3 millimeters across have been found at the **Blue Hammer showing** west of the Kootenay River (Brown 2004). The beryl occurs in granite pegmatite dikes that cut aplite dikes and quartz monzonite of the White Creek batholith. Brown (2004) also described several ice-blue aquamarine crystals up to 6 millimeters in diameter from the **OMG claims** west of Kootenay Lake. The crystals occur in quartz veins extending outwards from the quartz cores of pegmatite dikes into surrounding aplite and/or sedimentary rocks.

Legun (2004) compiled information on all known beryl occurrences in British Columbia and produced a map showing a "beryl belt" running the length of the province and into the Yukon Territory (map page 43).

Beryl Genesis

Hart and Lewis (2003) compiled information on the 10 documented beryl occurrences in the Yukon and western Northwest Territories in order to look for common characteristics. They found that the 10 occurrences are associated with peraluminous, alkalic, and metaluminous (?) granitoids from at least 6 different plutonic suites. Their findings were contrary to our expectation that most beryl would be associated with peraluminous granites. The results of the Hart and Lewis survey, however, do agree with the assertion by London

Left: *Brad Wilson shows off a 135-gram topaz crystal collected from a pegmatite pocket on Mt. Foster in 1993.*

Below: *Some of the gemmy beryl crystals found in several Mt. Foster pockets; the largest weighs 2.99 ct; collection and photo B. Wilson*

and Evensen (2002) that "no one granite source rock is uniquely or specially predetermined to produce beryl-bearing pegmatites." Future work on beryl occurrences not associated with peraluminous granites might help shed light on beryl genesis in general.

There is also the question of why the beryllium contents of the (presumed) parental granitoids at Lened, Regal Ridge, and True Blue are not exceptional. According to London and Evensen (2002), "It may seem logical that, as likely sources of the beryl-rich pegmatitic melts, these granites are themselves depleted in beryllium and other incompatible components."

Exploration

Recent discoveries of emerald in northwestern Canada have led to increased exploration expenditures; in 2003 approximately 30 percent of exploration expenditures (more than 3.5 million Canadian dollars) in the Yukon Territory were allocated to emerald exploration, mostly at True North Gem Inc.'s Regal Ridge property. It is not yet known if any of the existing properties will become producers, but given the size of the Canadian land mass, the low population density, the diverse geology, and the prior low level of exploration for colored gems, there will almost certainly be more discoveries.

Those interested in exploring for gem beryl should begin with a thorough literature search, keeping in mind the geology of Regal Ridge, Lened, and True Blue. Good places to start are Mulligan (1968) and Walton (2004); the latter in particular contains a detailed compendium of material from the literature concerning gem beryl.

Bill Wengzynowski uses a diamond saw to extract aquamarine crystals at the True Blue property. Photo by the author

Information on specific occurrences can be found in the MINFILE and NORMIN databases.

Examination of regional geology suggests that Regal Ridge was once located in the area north of Dawson City; therefore the area south of the fault in the Dawson area should be prospective for beryl, as evidenced by the Pluto occurrence. It is interesting to note that the Regal Ridge and True Blue properties are approximately 14 kilometers northeast and 30 kilometers southwest of the Tintina Fault, respectively, which might indicate a genetic connection.

There is also a cluster of beryl occurrences in the south-central Yukon and north-central British Columbia, which could be thought of as defining a beryllium-rich area.

Murphy et al. (2002) plotted beryllium reservoirs (plutonic rocks) and chromium and vanadium reservoirs (ultramafic rocks, volcanic rocks, and the Earn Group) in the Yukon Territory and suggested that the best place to look for emerald is where the two come together. Murphy et al. (2002) also plotted Regional Geochemical Survey (RGS) data for the Yukon Territory; these show high chromium near Dawson City and high vanadium close to the Yukon-NWT border (across from the Lened property), corresponding to mafic volcanics and Earn Group rocks, respectively.

Hart and Lewis (2003) noted that for an intrusion to develop enough beryllium to form beryl it must be "ultrafractionated." They observed that beryl occurrences in the Yukon Territory are almost all intrusion related. Geochemical analysis can help pinpoint fractionated pegmatites with elevated gemstone potential. In the Yukon, Hart and Lewis (2003) list the following intrusion-related targets: Old Cabin Creek, the Anvil batholith, Nisutlin South, the Nisling Range alaskite, Klondike, Fiddler, and the Tombstone Plutonic Suite. There are metamorphic targets in the Stewart River area, shearzone targets in the Klondike region, and black shale targets throughout the Selwyn Basin.

Companies exploring for emerald in northwestern Canada have been relying heavily on beryllium and chromium stream-sediment and soil samples. It will be interesting to see if this approach results in any new discoveries. What about berylometers? The weights, counting times, and detection limits likely limit their application to property studies at best. At the end of the day we can use geology and geochemistry to decide where to look, but careful prospecting will most likely be the most effective tool in discovering new occurrences.

Gem Beryl from Russia and Ukraine

Peter Lyckberg, currently of Luxembourg, is a nuclear inspector for the EU. He has extensively researched and collected the gem pegmatites of the former Soviet Union.

Russia

The most important areas in Russia for beryl are the emerald mines near Malysheva and the Alabashka pegmatites near Murzinka, both in the Urals; as well as Sherlova Gora and Adun-Chilon in Transbaikal. Numerous other deposits throughout Russia have produced interesting beryl specimens, but all pale in significance to the above mentioned districts.

Alabashka Pegmatite District

Steep-dipping, miarolitic pegmatites characterize the Alabashka, Murzinka, Lipovka, Sarapulka, Shaitanka and Aduy pegmatite fields of the central Urals. These fields are often confusingly and imprecisely referred to as "Murzinka," as all are situated near the village of Murzinka. In addition to being the name of a village, the term "Murzinka" refers to one of the lesser beryl-producing pegmatite fields in the area as well as to the intruding granite. The western endo and exocontact of the Murzinka-Aduy intrusions carry world famous specimens of topaz, beryl and tourmaline. The very best beryl and blue topaz come from the Alabashka pegmatite field, 5 to 8 kilometers north of Murzinka village.

Colored stones were first found near Murzinka in 1668. Soon after, miners began to work the Alabashka pegmatite fields for gems including beryl, topaz and smoky quartz. These stones were brought to Ekaterinburg to be cut and polished. When the author first visited the area in 1992, it had been virtually closed to Westerners since Swedish mineralogist Gustav Flink visited in 1916; George F. Kunz had toured the area the year before.

Alabashka pegmatites have produced some of the finest, sharpest gem beryl crystals in the world. Colors are often gold to yellow to yellow-green, less frequently green and rarely blue-green. Crystals typically range from 1 to 3 centimeters in diameter and 3 to 12 centimeters in length, in rare cases reaching lengths of 19 to 28 centimeters. Crystals are most commonly found loose in pockets, but several exceptional specimens have been found on matrix; some of these were recovered between 1990 and 1993 in the Staraya Mylnitsa and Kazionnitsa mines.

The **Staraya Mylnitsa mine** produces exceptionally fine heliodor, which occurs in association with albite balls that are sprinkled with tourmaline (dravite-foitite). Pocket No.133, which was encountered in 1990 at a depth of 13 meters, yielded 2.5 kilograms of beautiful specimens. The finest of these were faceted for jewelry and sold for next to nothing. Staraya Mylnitsa has also yielded fine schorl.

The last serious mining effort at **Kazionnitsa** ended in November 1993, but not before producing a few exceptional green beryl crystals to 23.5 centimeters. The best of these was a 7 centimeters flawless, "electric" green crystal. Gem golden beryl and foitite have also been recovered from Kazionnitsa.

All of the pegmatites around Murzinka are steeply-dipping veins, striking north-north-west. Veins are typically just a few meters wide (rarely reaching 30 meters) and 50 to 1500 meters long. Two exceptions are the flat lying Golodnaya vein and the main body of the Mokrusha vein, the latter being up to 150 meters wide, more than 500 meters long and reaching a thickness of 12 meters.

This 7.8 cm high pale yellow beryl on matrix is from the area around Mursinka in the Central Urals.

A. E. Fersman museum collection; Jeff Scovil photo

The **Mokrusha** vein in the Alabashka field produces world famous Russian blue topaz. The upper pocket zone contains cavities rich with pale to sky blue topaz mineralization. Blue topaz from Mokrusha in the Alabashka field sometimes forms attractive associations with smoky quartz and albite. On rare occasions, topaz occurs in the same pockets as beryl. Generally, however, it is only the lower pocket zone that contains beryl, primarily in the form of golden beryl, sometimes

*The **above** 15.6 cm long beryl from the Kazionnitsa mine is green under incandescent light and blue in daylight.*

Van Scriver specimen; Jeff Scovil photo

in association with multicolor tourmaline. Less frequently, beryl from these pockets is green or colorless.

As recently as the 1980s, Mokrusha was producing gem-quality deep golden beryl. In 1976 while doing exploratory work in the open pit, Sergej Borchov collected a superb, flawless, 234-gram deep golden crystal.

Golodnaya is an extension of the Mokrusha pegmatite that was offset by a fault. Beryl from both veins forms deep golden, sharp hexagonal crystals, which sometimes exhibit etched, pointed terminations. In 1985-86, large pockets of golden beryl with albite and smoky quartz (to 50 kilograms) were uncovered at depths of 12 to 22 meters in the Golodnaya vein.

Arsenopyrite and cosalite have been found as fine intergrowths in some beryl from Mokrusha (Popova et al. 2002). In fact, Mokrusha is rich in associated species that include hambergite twins to 5 centimeters, sharp purplish and pink fluorapatite and very fine, sometimes twinned microcline crystals.

Lesser-known veins in the Alabashka field, such as **Tyazhelovesnitsa** and **Startseva Yama** have also produced gem yellow beryl. Startseva Yama is famous for having produced, in 1828, a 2.5-kilogram, gemmy, beryl crystal (Svirina, 1994). Until 1991, Startseva Yama was exploited through 1.5 to 5-meter deep open pits, and fine golden beryl has been found loose in the soil only 1.5 meters from the surface. In addition from 1989 to 1991, underground mining was being carried out at the 12 to 27-meter level, where golden beryl up to 9 centimeters was found in association with albite, smoky quartz, schorl and muscovite.

The finest beryl specimens from the Alabashka pegmatite field can be found in private and public collections, including those at the Gorniy (Mining) Institute in Ekaterinburg and St. Petersburg, at St. Petersburg University, at the Fersman and the Vernadsky State Geological museums in Moscow, at the Natural History Museum in London and at the Ecole de Mines in Paris.

Izumrudnye Kopi

In the area around Malysheva, situated about 60 kilometers east-northeast of Ekaterinburg, desilicated pegmatites are surrounded by the famous emerald-alexandrite-phenakite mineralizations. The area was originally known as "Tokojava" for the stream that cuts through the region, and Tokojava is noted on many old labels on specimens from this region. Today the emerald-alexandrite district is more commonly known as Izumrudnye Kopi (meaning Emerald mines) or as Malysheva, the name of a nearby village founded for the sole purpose of supporting the mining industry.

The Izumrudnye Kopi deposits were exploited through the **Sretenskiy** and later through the Mariinskiy mine, located near Malysheva. The very finest alexandrite and emerald were found in the early days of mining and came from near-surface workings right by the Tokojava stream and from the Srentenkiy mine.

The phlogopite schists around Malysheva, the type locality for phenakite and clinobehoite, have been producing quantities of emerald since their discovery by peasant Maxim Kozhevnikov, who noticed crystals in the roots of an overturned tree. Kozhevnikov sold the stones in Ekaterinburg, where a gem-cutting industry was well-established. The alexandrite variety of chrysoberyl was also first found at Takovaya. Both alexandrite and emerald derive their intense color from trace amounts of chromium.

In 1995, an open pit at the old Sretenskiy mine was developed, but low production halted the mining operation. While it was in production, the Sretenskiy mine occasionally produced rare combinations of phenakite or alexandrite with emerald.

At the **Mariinskiy mine**, beryl occurs as emerald in different matrices: the palest green crystals are found in albite veins; light green to deep emerald green gems are found in the phlogopite sections; and small, deep green transparent crystals occur in talc and actinolite veins. Emerald crystals are commonly 1 to 3

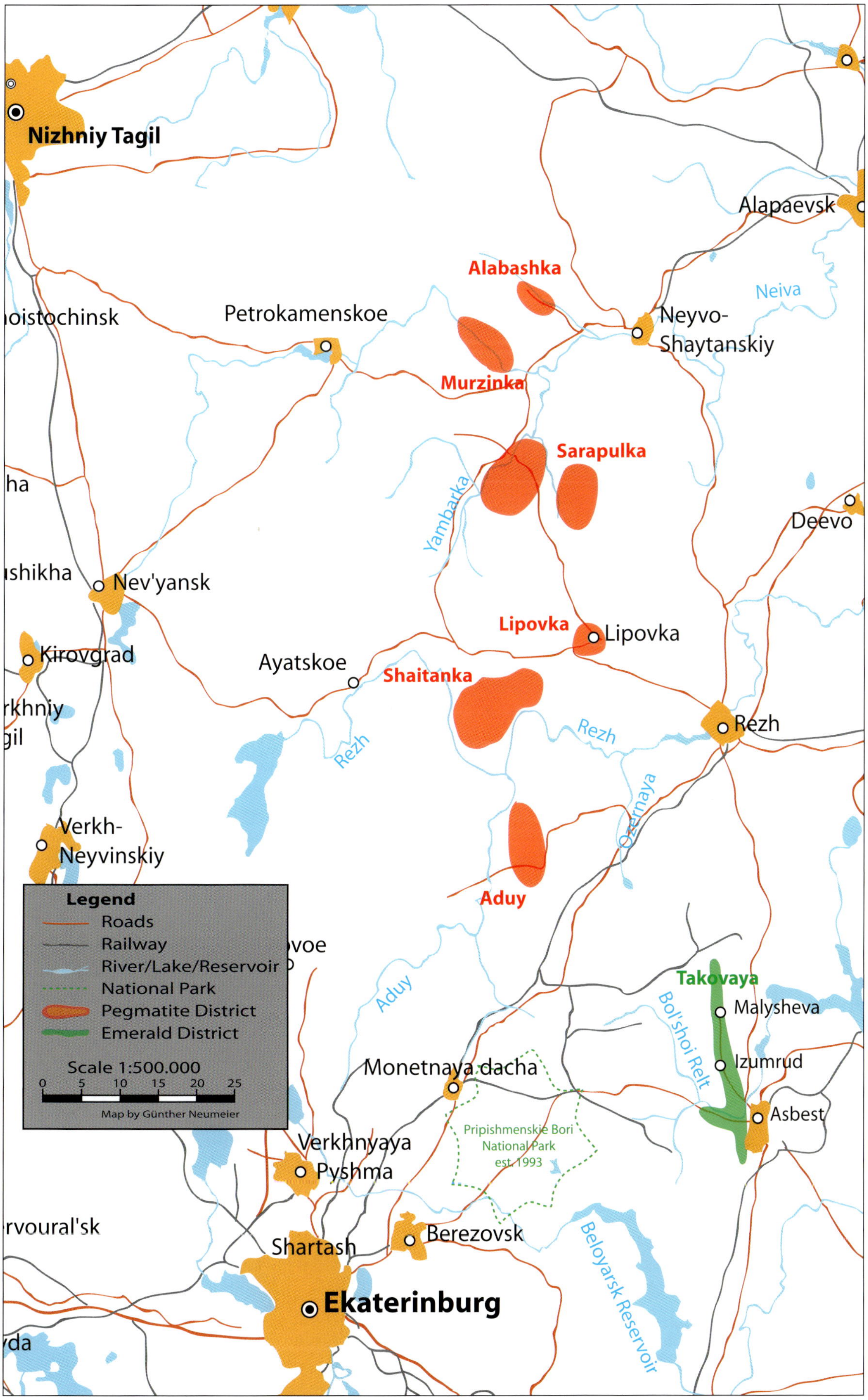
Nizhniy Tagil
Alapaevsk
Alabashka
Neiva
Neyvo-
Shaytanskiy
Petrokamenskoe
Murzinka
Sarapulka
Yambarka
Deevo
Nev'yansk
Lipovka
Lipovka
Kirovgrad
Ayatskoe
Shaitanka
Rezh
Rezh
Rezh
Verkh-
Neyvinskiy
Ozernaya
Aduy
Aduy
Takovaya
Malysheva
Bol'shoi Relt
Izumrud
Asbest
Monetnaya dacha
Pripishmenskie Bori
National Park
est. 1993
Verkhnyaya
Pyshma
Shartash
Berezovsk
Beloyarsk Reservoir
Ekaterinburg
Legend
Roads
Railway
River/Lake/Reservoir
National Park
Pegmatite District
Emerald District
Scale 1:500.000
0 5 10 15 20 25
Map by Günther Neumeier

A 5.5 cm high heliodor on smoky quartz from the Mursinka area Collection and photo Peter Huber

centimeters in length, with several Mariinskiy crystals reaching 20 centimeters. Larger crystals sometimes have a deep green transparent core, and a few exhibit banding that runs perpendicular to the *c*-axis. In a few rare pockets, colorless, light blue, white and yellow beryl crystals to 8 by 15 centimeters have been encountered. One of the finest specimens of Tokovaya emerald is a roughly 35-centimeter-wide spray of crystals in a matrix of phlogopite schist. The specimen is known as the "Leuchtenberg Emerald" and is part of the Bavarian State collection, which is housed at the Reich der Kristalle Museum in Munich, Germany.

As production costs rose to the point of being prohibitive, 1996-97 saw the end of most of the region's emerald mining; still, regional emerald reserves remain quite extensive, as emerald has been encountered in drill cores at depths of more than 1,000 meters (Laskovenkov, personal communication 1992).

Other Beryl Localities in the Urals

Two different types of gem-bearing pegmatites are prevalent in the **Murzinka** field. One type occurs northwest of the village where a dozen or so veins are enclosed in serpentinites that contain miarolitic pegmatites. Colored tourmaline and cordierite populate the veins. The other type, centered around the village, contain green and yellow beryl as well as clear, yellow or light blue topaz.

Similarly, two types of pegmatites characterize the **Sarapulka** field. Pegmatites east of the village of Sarapulka such as **Zolotaya Gora**, carry green and blue beryl crystals to lengths of 15 centimeters, some of which are doubly terminated. West of Sarapulka, irregular thin twisting veins carry red tourmaline and rhodizite as well as pink beryl on albite. Tabular pink beryl from this locality is typically 5 to 8 centimeters in diameter. The **Ministerskaya Yama mine** near Sarapulka and the **Mor's pit** near Shaitanka are the type localities for rhodizite. Nearby, the **Yuzhakova mine** produced rounded green beryl crystals that were often doubly terminated.

Generally situated north of a bend in the Aduy River, the **Aduy** field is the southernmost group of pegmatites in the central Urals. The **Semenikhinskaya** or **Monetnaya Dacha mine** produced two major pockets of large, gemmy green beryl crystals. One historic pocket, emptied over two weeks in 1899-1900, produced more than 500 kilograms of gem crystals (Fersman, 1925). A second pocket containing large quantities of beryl was opened in the 1930s (Alexander Rosskov, personal communication, June 1992).

Aduy beryl is gemmy, green and hexagonal, often exhibiting heavily etched terminations. Prism faces are also sometimes etched, with crystals ranging from 5 to 25 centimeters in length and 3 to 10 centimeters in diameter. The Fersman Museum in Moscow houses an impressive suite of mineral specimens from Aduy, whose more or less exhausted veins have not been mined since 1991.

In the southern Urals, more than 200 pegmatites are found in two major regions and in several smaller ones. The southernmost region around Miass in the **Ilmen Mountains** was heavily mined in the 19th century. The region primarily produced clear topaz crystals with complex forms weighing up to 3 to 4 kilograms. Green beryl crystals to 25 centimeters in length were occasionally found in miarolitic pegmatites; these crystals were sometimes associated with amazonite. The pegmatites are almost completely exhausted, and little remains in this area today.

At **Svetloe**, south of Miass and west of the Borisov granite intrusion, kaolin quarrying has exposed remnants of pocket-bearing pegmatites in which only the pocket walls and their contents have not been altered to kaolin (I. Pekov, personal communication). These pockets contain translucent, blue and gemmy golden beryl crystals that commonly reach 2 to 6 centimeters in diameter and 3 to 10 centimeters in length.

A 17.4 cm beryl from Urulga and a 20.2 cm long aquamarine from Sherlova Gora

Peter Huber collection and photo

Transbaikal

The Transbaikal region is home to another group of Russia's classic pegmatite and greisen deposits including **Adun Chilon**, **Sherlova Gora**, **Bortschovochniy Ridge** (Urulga, etc.) and **Soktuy**, not to mention more recently discovered pegmatite fields such as **Malkhan**.

Adun Chilon is the name of both a large grained porphyritic granite intrusion as well as the mountain built by this intrusion. Adun Chilon Mountain is littered with dozens of vertical pipelike pegmatites, 0.5 to a few meters in diameter. These are commonly mined to depths of 12 to 20 meters, with shafts sometimes reaching depths of 34 meters.

The deposit is believed to have been mined since antiquity and was rediscovered in 1829. State prisoners began mining Adun Chilon in around 1850. The unweathered pegmatites at Adun Chilon can produce sharp, yellow-green and green hexagonal gemmy beryl crystals to 30 centimeters. One rare sharp, blue translucent crystal is exhibited at the Fersman Museum in Moscow alongside a razor-sharp green gem beryl specimen with a steep termination.

Some crystals have been etched so deeply that their hexagonal shape is no longer obvious, while hexagonal crystals from some of the pockets exhibit typical etching along the edges. In one shaft, smoky quartz crystals weighing as much as 240 kilograms have been found in association with tourmaline and large yellow or blue gem wedge shaped topaz crystals to 4 kilograms.

The Adun Chilon pegmatites are not easily spotted through the meter-high grasses that cover the alluvial hills and valleys; thus, many of the specimens on public display, even those in Russia, labeled "Adun Chilon" are in fact from the physically close but genetically different deposit of **Sherlova Gora**. The Adun Chilon pegmatites are located in several valleys and along the ridges just south of a Buryat church known as Zagan-Obo and along Kuku-Sirken ridge. In fact, Kuku-Sirken is often seen on old labels that accompany specimens from Adun Chilon.

The Sherlova Gora greisen deposit probably contains the largest reserves of easily extracted gem beryl in Russia. Gem blue beryl crystals were discovered in 1723 at Sherlova Gora, which at that time was considered as part of the Adun Chilon Mountains. Today a strong geographic and genetic distinction is properly made between the two deposits.

This greisen deposit produces beryl crystals that are typically heavily striated and color zoned perpendicular to the *c*-axis. They commonly appear cylindrical in shape due to the presence of di and tri-hexagonal faces. Terminations are generally simple pinacoids. Sherlova Gora beryl crystals have been described in the literature as having complex terminations but the source of many of the described crystals is actually Adun Chilon.

This 8 cm high aquamarine from Adun Chilon is housed in the Fersman Museum in Moscow.

Jeff Scovil photo

Sherlova Gora beryl is green-blue, blue, green and rarely yellow-green or true golden heliodor. The largest known preserved aquamarine crystal from the locality is 31 by 5 centimeters and is on exhibit at the Natural History Museum in London. In most of Europe's large museum collections, there is a suite of Sherlova Gora beryl, which is sometimes labeled the "The Urals."

Over the last two to three decades, the **Orlovskoe** "apo granite" has been mined for tantalum that is found in the vertically fractionated albitized granite and in greisen deposits that contain amazonite pegmatite veins. Some zones at Orlovskoe are heavily enriched with topaz that occurs as clear small crystals to a few centimeters in open cavities. Gem golden beryl crystals to 25 centimeters have been encountered in clay filled pockets.

In the past, minerals from the Transbaikal region would find their way into the old gem and mineral trading centers of the Urals, and some old collections still contain Transbaikal specimens labeled "The Urals."

Situated southeast of Lake Baikal and just to the north of the Chikoy River, the Malkan Mountains are home to at least 50 pegmatites, some rich in pink, green and yellow tourmaline. Beryl occurs in the **Danburitovaya vein** as white to tan terminated, hexagonal crystals frozen in matrix.

The **Vodorazdelniy mine** in the Menza district, southwest of the Malkhan Mountains is a miarolitic pegmatite that was mined for piezoelectric quartz. Byproducts of the mining effort, flawless bicolor tourmaline pencils (sometimes on matrix), green gem dravite and pink to colorless beryl "floaters" were thrown on the dumps. Beryl at the Vodorazdelniy mine can be bicolor (clear or slightly light green to orange-pink), but is often a salmon color. Crystals are either heavily etched, or they occur as slightly etched, large hexagonal tabular crystals 4 to 15 centimeters in diameter and up to 4 centimeters thick. Beryl is found in association with albite, lepidolite, muscovite and spodumene.

Pegmatites of the **Bortschovochniy Ridge**, rich in tourmaline, topaz and beryl, were first discovered and mined in the 1900s. The **Urgula** pegmatites in the Bortschovochniy Ridge produced some of the early 20th century's finest champagne-colored topaz and green beryl.

There have been few modern attempts to mine these veins, though the 300-meter-long **Savateevka** pegmatite was exhausted in a recent spodumene recovery effort. According to Fersman (1925), before being mined out, the Savateevka pegmatite produced multicolor tourmaline to 3.5 kilograms, lepidolite, quartz crystals, albite, apatite, garnet and pink beryl. Other primary veins in the region are the **Mokhovaya**, **Gremyachenskaya** and **Suchja Pad'**, but none can boast a notable production.

Other Beryl Deposits in Russia

Gem golden beryl crystals have been found in very rare pockets in the muscovite pegmatites of the **Mama district** northeast of **Lake Baikal**. These crystals sometimes reach 15 centimeters in length and are associated with smoky quartz. According to a handful of collectors from Irkutsk who have seen samples from the area, these fine yellow to golden specimens are sharp, hexagonal

***Facing Page**: Nine of the fourteen beryl specimens depicted on Plate 44 of Reinhard Brauns' Das Mineralreich are from Russia. A few carry inaccurate locality designations.*

***1**: Emerald in mica schist, Habachtal, Austria; **2**: Emerald, Muzo, Colombia; **3**: Emerald group, labeled Tokojava (now known as Izumrudnye Kopi); **4**: Emerald in calcite, Muzo, Colombia; **5**: Beryl, Shaitanka, Urals; **6**: Golden beryl, labeled Murzinka, Ural (more precisely Alabashka pegmatite field); **7**: Beryl, Murzinka; **8**: Aquamarine, Nertshinsk, Transbaikal; **9**: Aquamarine, Adun-Chilon, Transbaikal; **10**: Blue beryl, Slieve Corra, Ireland; **11**: Blue beryl, labeled Adun-Chilon (from the pegmatite now known as Sherlova Gora); **12**: Beryl, Bortschovochniy Ridge, Transbaikal; **13**: Aquamarine cut stone; **14**: Aquamarine labeled Adun-Chilon (from the pegmatite now known as Sherlova Gora)*

Plate Number 44: Beryl

from Reinhard Brauns' *Das Mineralreich* (The Mineral Kingdom); Stuttgart, 1903

(the mineral descriptions can be found on the facing page)

and gemmy with pinacoid terminations.

Russia has innumerable other pegmatites, many of which have produced at least some gem beryl crystals. The **Polimineralnaya vein** in the **Zolotaya Gora** (translated Gold Mountain) district in eastern Transbaikal, for example, contains a rich mineralization of albite, schorl, lepidolite, petalite, amblygonite, cassiterite, Ta-Nb minerals and mica. The more than 150 pockets encountered in the district have produced multi-color tourmaline as well as colorless and pink beryl. The Svetloe Tungsten greisen deposit near **Chukotka** in northern Siberia has likewise produced ice blue beryl crystals up to 8 centimeters long.

Ukraine

In 1867, Ossovski first found topaz, beryl and smoky quartz in a glacial drift in Zhitomir oblast (county). These finds led him to the world famous **Volodarsk-Volynskiy** pegmatites. The region was under study in 1930, and by 1931 the pegmatites were being mined. During the Cold War, Volodarsk-Volynskiy, along with some 1,500 pegmatites in Kazakhstan, was being heavily exploited to supply the Soviet Army with piezoelectric quartz. Little was known about these deposits in the West until the mid-1980s when gem beryl began to flow, first to Europe's mineral shows and then to those in the United States. Though nearly 100 different mineral species were found in Volodarsk-Volynskiy, only a few of the pegmatites yielded collectable gem species such as topaz or beryl. Of the region's hundreds of pegmatites, only about 2 percent carry beryl.

At Volodarsk-Volynskiy, gem crystals of topaz and much more rarely gem beryl are confined to oval chamber pegmatites typically 4 to 15 meters in diameter and 6 to 25 meters in length. Large pegmatites can reach 15 meters in diameter and up to around 70 meters in length, with pockets occasionally reaching 30 meters (Koshil et al. 1991). Situated along the west contact of the Volodarsk-Volynskiy-Volhynsk gabbro anorthosite pluton on the Ukrainian craton, these miarolitic pegmatites are in a 22-kilometer long, 500 to 1,500-meter wide north to south trending belt (Koshil et al. 1991). Veins were located by grid drilling every 50 meters over an extensive area to depths of 100 to 150 meters.

Beryl crystals from Volodarsk-Volynskiy are often heavily etched and range in color from dark olive green to the more common yellowish green. Numerous large crystals have been found, some weighing up to several kilograms and measuring 20 centimeters in diameter, although most of the crystals are between 1 and 3 centimeters in diameter and 3 to 10 centimeters in length.

The Volodarsk-Volynskiy pegmatites are accessed through 6 main shafts that are connected by tunnels at the 50, 100 and 150-meter levels (Koshil et al. 1991). Dozens of pegmatites were mined on each level until mining was halted in 1996. It is very unlikely that these mines will ever reopen, though local collectors continue to search the open pits for pockets and gem material.

The main component of the larger pockets at Volodarsk-Volynskiy is quartz, which often forms crystals up to a meter in length. The quartz is associated with mica, kaolin, microcline and albite. Pockets also contain broken and rehealed shards of quartz and a suite of other minerals including SiO_2 in the form of opal. Roughly 10 percent of the pockets had some topaz, and about 2 percent carried beryl. Though rare, beryl-bearing pockets can be quite extraordinary. In 1992 for example, a single 12x4x5-meter pocket delivered over one ton of beryl!

Other Ukrainian Deposits

In addition to Volodarsk-Volynskiy, there are a few lesser-known localities that are or have been worked for beryl. The **Krutaya Balka** deposit, near the western shore of the Sea of Azov near the city of Mariupol, for example, has yielded a find of very pure emerald crystals to 2 centimeters. These were found in a schistose phlogopite that is geologically similar to the Tokovaja deposit, though the crystals and quantities of Krutaya Balka emerald have been miniscule in comparison to those recovered from Tokovaja.

The many classic and emerging localities for Russian and Ukrainian beryl create a wonderful challenge for any collector wishing to build a representative suite of Russian and Ukrainian beryl specimens.

Acknowledgements:

Building experience in and knowledge of any mineralogical region requires more than even the most committed efforts of any one person. It requires the cooperation and assistance of countless people who act as companions and mentors, who share their expertise and experience, and who sometimes risk their own security to help a fellow collector and student. The people who have generously supported my study of and travel to the Russian gem deposits are too numerous to mention; still, I owe each a debt of gratitude, and I sincerely hope that sharing here a fraction of the information that they have helped me to gather will stand as some small tribute to their efforts.

Beryl
16 cm
Volodarsk-Volynskiy, Ukraine
Wayne Thompson collection; Jeff Scovil photo

Mellow Yellow:

Michael Wise has been a geologist on the curatorial staff of the Smithsonian Institution since 1988. His research focuses on pegmatite mineralogy.

Beryls are some of the world's most popular gemstones, largely because of their gorgeous colors. Emerald and aquamarine are easily the most recognizable members of the group and are without a doubt the most popular. Yellow-colored beryl, by comparison, is arguably the least desirable of all colors and is generally ignored by most consumers. Emerald and aquamarine are more salable to the consumer; thus it is not surprising that the removal of the yellow component from emerald to enhance its color and from green beryl to produce aquamarine is common practice in the gem industry.

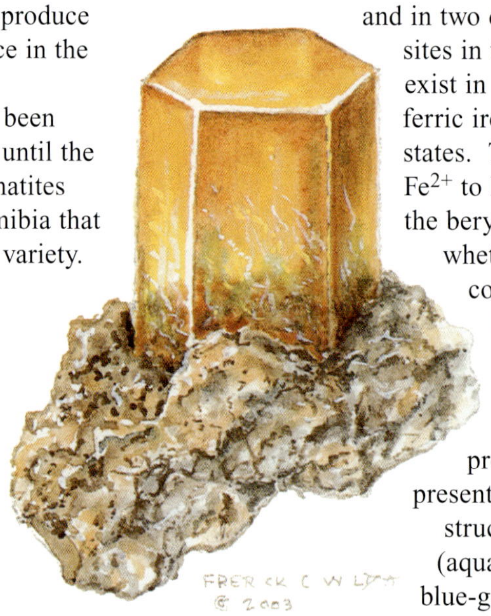

Massachusetts artist Fred Wilda rendered the watercolor of a gem beryl crystal that was found in 1974 in the Palermo No. 1 mine in Groton, New Hampshire. Collection Bob Whitmore

Although yellow beryl has been known for centuries, it wasn't until the discovery of the Rössing pegmatites near Klein Spitzkoppje in Namibia that yellow beryl became a named variety. In 1910, golden yellow to brownish yellow crystals were found in complex granitic pegmatites and were claimed to have properties different from other known finds of yellow beryl. The Rössing material allegedly displayed blue phosphorescence, opalescence, a weak green fluorescence, weak radioactivity and an alexandrite effect. To increase the popularity of this new find, the beryl was called "heliodor." The name was derived from the Greek terms "doron" meaning gift and "helios" meaning sun, i.e., gift from the sun. In spite of the fact that claims about the special properties of the Rössing material were proven false, the term "heliodor" remained in use. Today it is commonly used to describe gem-quality beryl that may occur in a wide range of hues from the yellow palette. These include pale yellow to golden yellow, orange-yellow, yellow-green, and brown-yellow. This color range has led to an inconsistent use of the name "heliodor." The term is sometimes expanded to include, for example, all of the stones produced by an accepted heliodor deposit. Beryl from the Volodarsk-Volynskiy pegmatites in the Ukraine (see page 55), for example, ranges in color from golden to yellow to decidedly green; however, all of the stones from this deposit regardless of color are often referred to as heliodor.

Most of the yellow in beryl is due to trace amounts of iron residing in the beryl structure. The iron content of yellow beryl varies from about 0.09 to 1.8 weight percent. Some yellow beryl reportedly contains trace amounts of uranium, which is suspected to contribute to its color.

Extensive research on the color of beryl indicates that iron can occur in two valence states and in two different crystallographic sites in the beryl structure. Iron may exist in beryl as ferrous iron (Fe^{2+}), ferric iron (Fe^{3+}) or as both valence states. The relative concentration of Fe^{2+} to Fe^{3+} and their position within the beryl structure largely determines whether yellow, blue or green colors will develop. When Fe^{3+} substitutes for aluminum in the octahedral position in the beryl structure, a yellow color is produced. If, however, Fe^{2+} is present in the channels of the structure, a blue color (aquamarine) usually develops. A blue-green color is generated if both types of iron are present. If, on the other hand, Fe^{2+} resides only in the octahedral position or Fe^{3+} only in the channels, then the beryl exhibits no color.

The effect of heating and irradiation on the color of iron-bearing beryls is well documented. Heating heliodor or golden beryl to approximately 250-350 degrees Celsius causes a reduction of Fe^{3+} to Fe^{2+}, thereby eliminating the yellow color and producing colorless beryl. Further heating to temperatures around 430-475 degrees Celsius causes a change from pale to dark blue aquamarine. Irradiation can reverse this process restoring the yellow color. Some irradiated beryls become a bright lemony yellow, while others become a more golden color. Some colorless beryl (goshenite) containing minor amounts of iron can be changed to yellow beryl by radiation. For example, colorless beryl from Orissa, India with 0.5 to 0.7 weight percent iron can be changed to yellow with irradiation and

Heliodor and Golden Beryl

subsequent heating to 300 degrees Celsius.

Although yellow beryl is common in pegmatites worldwide, most is not of gem quality. Pale yellow and golden beryl is typically characteristic of poorly to moderately mineralized

Left: *Gem rough from the Baboon Hill mine in Zimbabwe's Mwami district, the largest stone is 1.5 cm in diameter*

Photo Jan Kanis

Right: *This 8 cm high golden beryl crystal from the Mursinka area in Russia's Ural Mountains is housed in the Houston Museum of Natural Science.*

Photo Jeff Scovil

Heliodor's yellow color may be induced by natural radiation emanating from nearby radioactive minerals such as uraninite, U-bearing columbite-tantalite, zircon, etc. The gamma, beta and alpha particles emitted from such minerals can ionize iron in the beryl structure causing a change from mainly divalent iron (Fe $^{2+}$*), which imparts blue, to a mixture of divalent and trivalent iron (Fe*$^{3+}$*), imparting green. Further irradiation causes even more of the iron to be converted to Fe*$^{3+}$ *(imparting a yellow to orange color).*

Jim Nizamoff describes having observed this phenomenon at the Palermo #1 mine in New Hampshire. "In the fall of 2003 we encountered a large (35 to 40 centimeter long) beryl crystal hosted in triphylite. The top of the crystal was about 10 centimeters in diameter and was blue-green. As we excavated down to the middle part of the crystal, its color became greener. Toward the bottom, the color graduated to yellow then almost to orange. Within 10 centimeters of the bottom of this beryl, we found an area rich in uraninite. I surmise that the crystal had been irradiated to different degrees relative to its proximity to the uraninite, with the yellow-orange section enduring the most intense exposure. The surrounding feldspar was a brick red color that was likely also radiation induced."

Found in July 1999, the specimen pictured ***right*** *is from Maine's Orchard pit in Buckfield, Oxford County. Beryl in the Orchard pit is found in association with U-bearing ferrocolumbite and/or uraninite, both probable sources of radiation. The bottom of this 7-centimeters high crystal was closer to the source of the radiation than was the termination.*

Collection and photo by Orchard pit operator Gary Freeman

granitic pegmatites. The pegmatites are often simply zoned with the yellow beryl typically occurring in the outer zones of the body. Associated minerals may include almandine, columbite-tantalite, apatite, tourmaline and simple Fe-Mn phosphates. Heliodor occurs primarily in miarolitic cavities of anorogenic granites and their associated pegmatites. It is often etched and tends toward yellow-green in color. In some pegmatites, aquamarine may accompany yellow or common green beryl. Rarely do yellow beryl and emerald ever occur together. At the Crabtree emerald mine in Spruce Pine, North Carolina, the pegmatite carries bicolor emerald green and yellow beryl crystals; some crystals may show emerald rimming yellow beryl. Although some of the emerald from this locality is of gem-quality, the yellow beryl is not. Exceptional localities for gem-grade yellow beryl and heliodor are known from the Slocum quarry and Roebling mine, Connecticut (USA) (page 28); Luumäki, Finland (page 63); Minas Gerais, Brazil (page 84); Rössing, Namibia (page 78); Ikalamavony, Madagascar; Volodarsk-Volhynia, Ukraine (page 55); and the Alabashka and Mursinka pegmatite fields, Russia (page 49).

New Hampshire beryl collector and owner of the Palermo mine (Groton, New Hampshire) Bob Whitmore has been experimenting with enhancing the color of beryl crystals. The samples in the three pictures to the left have all undergone various enhancements.

The specimens depicted on the ***upper left*** *are all from Pakistan. The largest crystal was naturally light blue, the smaller ones were colorless. Bob heat-treated and irradiated the entire lot at 500 MRs, achieving various shades of yellow to gold.*

The beryl rough in the ***middle*** *photo is all from Bob's own Palermo mine. Palermo is known to produce natural yellow beryl, though these particular specimens came out of the ground colorless. All of the samples were heat treated and irradiated. The crystals in the top left and bottom right were irradiated at 200 MRs; those in the top right and bottom left at 1,000MRs.*

All of the specimens pictured on the ***lower left*** *originated in Afghanistan as colorless beryl. The samples were heated treated than irradiated at 400 MRs to give them their yellow color.*

All of the specimens are in the collection of Bob Whitmore.

Photos Jeff Scovil

The Mysterious Golden Water of Tajikistan

Dmitriy Belakovskiy is curator of the mineral collection at the Fersman Mineralogical Museum at the Russian Academy of Sciences in Moscow.

I first saw the golden beryl from Tajikistan in the spring of 1995, after the material had already been on the market for a while. An American dealer showed me a few decent specimens sitting on a matrix of albite, muscovite and quartz as well as a number of loose crystals. Looking at the matrix and shape of some of the samples, I was reminded of the many specimens I had seen over the years of Pakistani aquamarine. The dealer said that the specimens had been recently mined from somewhere in the eastern Pamir Mountains. No further information was known, but he assured me that he would try to get a more specific locality.

Beryl variety Heliodor
Both crystals are labeled as coming from Zelatoya vada, Rangkul, Tajikistan; the specimen on the left is 3.3 cm high, the specimen on the right is 3.6 cm high.
Private collection; Jeff Scovil photo

Though I had some knowledge of the eastern Pamir, I was unfamiliar with the material and could offer no help in determining the locality. I did, however, suggest that if these golden beryl specimens were from the eastern Pamir, they **may be** from the Rangkul area, as there is a well known gemstone area near Rangkul Lake about 60 kilometers east of the town of Murgab in Tajikistan.

I took a decent specimen in trade for the Fersman Museum collection, and upon returning to Russia, I showed the specimen to Anatoliy Skrigitil'—a geologist who had spent a number of years working in Tajikistan and Kyrgisia, particularly in the Rangkul area. Anatoliy had personally discovered a few gemstone localities in that region and was at the time my best source of information about the mineralogy, geology and gemstones of the eastern Pamir. To my surprise, Anatoliy was certain that heliodor had never been found in Rangkul or anywhere in the Pamir. Considering a potential source for the material, he mentioned that the pegmatites in Kyrgizia (north of Pamir) might have produced golden beryl.

In time, more material began to appear on the market, and to my dismay, the locality being given for them was the one that I had inaccurately suggested: Rangkul, Tajikistan! In addition, the names "Zolotaya vada" or "Zelatoya vada" were cited on some dealers' labels. Oddly, this mysterious Tajiki golden beryl locality now had a Russian-sounding name! "Zolotaya vada" is clearly a distortion of the Russian words «Золотая вода» meaning golden water; «Zelatoya vada» is «Желтая вода» meaning yellow water (the Russian words for gold and yellow have the same root). While I may have inadvertently started the Rangkul inaccuracy, I was grateful that the "golden water" designation had not come from me. This new, more specific locality heightened the mystery of the material's source. I began to poll the mineralogists and geologists that had been or were presently working in Tajikistan, but none had ever heard of a "golden water" mine or deposit.

"The improver of natural knowledge absolutely refuses to acknowledge authority, as such. For him, skepticism is the highest of duties; blind faith the one unpardonable sin."
-Thomas Henry Huxley (1825-95) British biologist and educator

On the world market, rumors were circulating

Both the 3 cm long beryl with schorl on feldspar pictured ***upper right*** *and the 2.7 cm long single crystal with spessartine pictured* ***below*** *are said to be from Zelatoya vada in Tajikistan's eastern Pamir Mountains. Both specimens are in private collections; the upper photo is by Stefan Weiss, the lower is by Jeff Scovil.*

that Tajiki golden beryl was nothing more than irradiated Pakistani aquamarine; still, more and more material was becoming available. To my amazement by the end of the 1990s, golden beryl from Tajikistan could be found among the stock of nearly every dealer carrying Pakistani and Afghani minerals, but the striking similarity between the matrices of Pakistan's aquamarines and Tajikistan's heliodors did not seem to faze most dealers or collectors (though the American dealer who first introduced me to the material did stop carrying it).

On the Russian front, not a single Tajiki heliodor crystal had appeared at the gem and mineral shows in Moscow or Saint Petersburg, in spite of the fact that other Tajiki gemstones were regularly turning up there. I continued to try to get more detailed information about the locality, as collectors often asked me about it. My suspicions were heightened as some traders became angry about my expressing doubts about the locality. One dealer once even offered a nice piece for the museum if I would confirm the Tajikistan origin of the material.

Rumors continued to circulate until a letter written by John White was published in the *Mineralogical Record* (White, 2000). In his letter, John reproached the doubters among us for sharing our concerns and "affecting the market for these legitimate specimens." Indeed, we failed in our attempts at the Fersman to confirm whether crystals were artificially or naturally irradiated. I agreed with John in that the specimens should be considered legitimate until proven otherwise; still I was puzzled about this material from Tajiki locality with a Russian name that was being sold by nearly everyone except for dealers from Russia and Tajikistan.

I have visited the Rangkul area and have seen first hand how difficult if not impossible it would be to hide mining activity on the barren landscape. In spite of my own experience, I began to believe that a hidden mine producing hundreds or even thousands of heliodor specimens really did exist in the Pamir, and that the secret was so closely guarded that even the local people had never heard its name, never seen the mine and had never seen a single piece that it had produced.

My lingering doubts about the authenticity of these specimens were unexpectedly and unequivocally confirmed, however, when I visited a small mineral show in Harrogate (Great Britain). There I noticed a table of the familiar golden beryl from Tajikistan. The seller assured me that he had personally collected the material, though he refused to provide details about the locality. The dealer claimed to have traveled there with the Minister of Geology of the Tajikistan Republic ("such a mustached guy"), whose name he had forgotten. Among the pieces on this dealer's table were a few, also labeled "Tajikistan," that were unlike any that I had seen at shows in the United States. These were golden beryl crystals on muscovite that exhibited a flattened, pinacoidal habit. Except for their color, these specimens looked exactly like goshenite crystals from the Ping Wu mine in China. It is truly a land of wonders, Tajikistan!

Finland's Famous Find

Native to Sweden, Peter Lyckberg, has been a lifelong mineral collector traveling to the ends of the earth to build his collection. He shares here a world-class beryl find made relatively close to his home.

Over the last two decades some of Europe's and even the world's finest colored, gem beryl has been unearthed from the deep forests of Scandinavia; the best having been found as recently as May 2004 near Luumäki, Finland.

That now famous deposit was first discovered in 1982 when a shard of clear, colorless topaz was picked up at a road construction site in the Kännätsalo (Finnish for Drunken Forest) on Kivi Järvi (Stone Lake), near Luumäki, Karelia in southern Finland. That shard got the ball rolling, and local gem enthusiasts and cutters began to lay claim to the granitic pegmatite running parallel to the road.

The deposit, which is situated on a small island in one of Finland's 70,000 lakes, subsequently produced significant crystals and gem material and was documented in an article by A. I. Lahti and K. A. Kinnunen that appeared in the Spring 1993 issue of *Gems & Gemology*.

The Luumäki pegmatite exhibits a large quartz core with some smaller offshoots. The quartz core often has multiple pockets running along it. Within the core, pockets can reach 4 meters. The core pockets typically only yield quartz, which is sometimes collectable or suitable for cutting.

In 1992, collector Risto Pitkänen reported observing one gem beryl-bearing pocket directly below a series of large opaque, common beryl crystals, some of which reached a meter in length. The rock near the bases of some of the common beryl hosts radioactive minerals such as monazite. In these areas, the common beryl had been dissolved, and flawless, gem beryl occurred in a hypergene, clay-filled pocket.

More recent discoveries at the **Karelia Beryl mine** differ from the earlier finds in that the more recent finds are not associated with a primary non-gem mineralization, although such a zone could have once existed above the new pocket, which had long since eroded. Larger gem-quality beryl crystals are sometimes overgrown with a secondary, pale, porous, yellowish beryl.

Since its discovery, the deposit has produced more than 110 kilograms of beryl. About 30 kilograms, nearly 30 percent of the beryl recovered, were suitable for faceting. Most of the facetable material was in the form of small (10-25 millimeter) fragments, though some of the rough was as large as 5-8 centimeters in diameter.

Some 10 kilograms of the material recovered had no visible inclusions, and in exceptional cases, these crystals weighed upwards of 1 kilogram apiece. These larger stones exhibit a fine green to yellowish green color. Crystals typically show rounded terminations, striated surfaces and are etched, though they retained their hexagonal shapes. Several thousand carats of cut stones could have been fashioned from the lot of larger crystals; most were instead preserved as specimens.

One 3 centimeter long golden beryl from the locality was recently studied and was shown to have dozens of parallel channels running perpendicular to the *c*-axis of the crystal. These channels were filled with rusty hypergene clay minerals.

The largest cut beryl from Karelia was faceted in Finland. In fact, Finnish master faceter Reimo Armas Römkä has cut a total of 90 to 100 large (25-50 carat) stones. In spring 2004, about 11,000 round brilliants, each 10 millimeters in diameter were cut in either China or in Idar-Oberstein. Most of the cut stones have been sold on the Finnish domestic market.

Though mining continued through the 2004 season, no further beryl pockets have been discovered.

"Elina," a 620 gram, 9.5 cm high, flawless green gem beryl recently found at Luumäki

Collection and photo Peter Lyckberg

China: An Emerging Source

Hunan native Guanghua Liu, presently lives in Tübingen, Germany. He is a coal geologist, mineral dealer and guest professor at the China University of Mining and Technology in Beijing.

A Yi girl poses before a karst landscape in Shilin, Yunnan China.

Photo taken in summer 2000 by the author

China boasts a large number of beryl-bearing deposits that produce a wide variety of beryl crystals. Gem-quality beryl, however, is not commonly found, and Chinese beryl has not yet gained particular attention in the collector market. That is not to say that China will never be known for gem beryl, only that China's gemstone potential, on the whole, is largely undeveloped. If the country's economy continues its breakneck growth and the Chinese continue to import western culture, the gem and mineral market in China is likely to expand rapidly, as did the markets in Japan and Hong Kong in the 1970s and 80s.

China's beryl crystals are found generally in the northwestern and southwestern provinces along the major mountain ranges. Most beryl deposits are associated with granites, pegmatites

Variety	Locations (Provinces)	Mineralization
Aquamarine	Xinjinag, Yunnan, Fujian, Hunan, Hubei, Shanxi, Hebei, Jiangxi, Sichuan, Inner Mongolia	Mainly granitic pegmatite, occasionally high temperature hydrothermal veins
Heliodor	Xinjinag, Yunnan	Mainly granitic pegmatite, occasionally high temperature hydrothermal veins
Emerald	Yunnan, Heilongjiang, Qinghai	Predominantly high temperature hydro-thermal veins, skarn
Goshenite	Kunagding, Danba, Sichuan	Predominantly high temperature hydro-thermal veins, some places may also be related to pegmatite
Morganite	Sichuan, Xinjiang	Predominantly high temperature hydro-thermal veins, some places may also be related to pegmatite

Known Chinese occurrences of some of the color varieties of beryl

# of World-Class Beryl

and hydrothermal veins. While a vast number of granites and pegmatites throughout China have been identified, they have not been adequately explored. A great many beryl deposits are thus likely awaiting discovery.

New Frontier

Xinjiang Uighur Autonomous Region (XUAR) is located in northwest China. The region encompasses an area of 1.6 million square kilometers (617,763 square miles)—one-sixth of China's total territory—but is home to only 1.4 percent of the country's population! Xinjiang, which in Chinese means "New Frontier," is bordered in the south and east by the Chinese provinces of Qinghai, Gansu and Tibet. In the west and north, the region borders the countries of India, Pakistan, Afghanistan, Tajikistan, Kyrgyzstan, Kazakhstan, Russia and Inner Mongolia.

More than 19 million people from 47 different ethnic groups populate Xinjiang. The Uighur form the majority of Xinjiang's mostly Muslim populace, but the population of Han Chinese in the region has grown rapidly over the last half century.

The Altai and Tian Shan Mountain Ranges in the north and Kunlun Mountain in the south form the province's geographic framework. In the north, the Junggar and Turpan basins separate these ranges. In the south, the Tarim basin, which lies across Tian Shan Mountain, developed a long inland river and lake. Large gaps in the surrounding mountains ranges expose the northwestern section of the Junggar basin to the climatic influences of Siberia. Junggar thus experiences cooler temperatures and more precipitation than do the more southern basins. Colder temperatures generally restrict the mining season in the Altai area from May to early October.

Xinjiang is dominated by a desert climate and is characterized by spectacular landscapes that include snow-capped peaks, petrified forests, and extensive deserts with huge dune structures. These environs are host to a great variety of plants and animals. The ancient Silk Road cut through Xinjiang, and brought a mix of eastern and western cultures to the region. These cultures left behind stunning relics that include ancient city ruins and decorated caves as well as historic temples and tombs.

This 8 cm high blue beryl on matrix is from an unnamed pegmatite in the Altai Mountains. Author's collection; Jeff Scovil photo

Xinjiang is China's most important gemstone-producing region and is particularly well known for its aquamarine. The major localities for gem beryl are distributed through the Altai Mountains, though notable localities are also found in the Junggar basin and the Tian Shan Mountains. Geologically, the beryl-bearing deposits in Xinjiang are of magmatic, pegmatitic or of high-temperature hydrothermal origin.

In the region surrounding the city of Altai, including the counties of Fuhai, Fuyun and Qinghe, there are thousands of pegmatite veins and granite intrusions. Beryl-producing mines

Four miners gather at the entrance to an unnamed pegmatite mine not far from the city of Altai. The photo was taken by the author in 1996.

and quarries are widespread and are visible in many places throughout the area.

The aquamarine and tourmaline deposits at **Jimokai** in Fuhai as well as the metal-mica-beryl deposit in **Ayoubulake** in Fuyun are both typical granitic pegmatite formations. Beryllium, niobium, cesium and rubidium minerals are well developed within these granites. In fact, more than 120 minerals including beryl, spodumene, tourmaline, garnet, topaz and rose quartz have been widely found in the area.

Although most of the region's beryl originates in the granitic pegmatites, aquamarine that formed during an episode of high temperature hydrothermal mineralization has reportedly been found in the **Qinghe** area.

Simple mining methods (i.e., hand tools and explosives) are employed throughout the many small mines in the Altai area. Pits thus tend to be less than about 100 meters deep. Beryl in the Altai area is generally found "frozen" in the host rock and forms mainly light blue, translucent to transparent crystals that range from 1 to 10 centimeters in width and from 3 to 30 centimeters in length. Chatoyant aquamarine, morganite and heliodor are occasionally found. Perfect crystals on matrix from these localities are rarely brought to market as the area miners tend to be interested solely in gem rough and not in mineral specimens. Miners break up many large crystal specimens in search of whatever gemmy sections they might contain.

In September 1999, I visited a small, unnamed mine in the Altai mountains situated in a small valley about 25 kilometers northwest of the city. The road leading to the mine was rocky and narrow—wide enough only for one tractor or Jeep to pass. The mine was only about 30 meters deep and advanced along a 2-meter wide quartz vein. Nearby, there were several abandoned pits, each some tens of meters wide.

The mine was owned by a group of Altai city residents, most of whom held positions in state-owned enterprises or in the government. The group worked as gem hunters during the summer months, on weekends and holidays. The mine owners reported having recovered about 50 kilograms of gem quality beryl by the time I visited in 1999, but have not reported any good finds of beryl since. As is typical for the region, beryl crystals are encased in the surrounding quartz, and all tend to be broken by the force of the explosives used to extract them.

Koktokay, the largest mine in the area, is located in the northeastern section of Fuyun county, 30 kilometers northeast of the town of Fuyun and 150 kilometers from the city of Altai. The area around Koktokay is rich with pegmatite veins. The Number 3 vein is the richest in rare metal and gem minerals and has been worked by the Chinese government for industrial rare minerals for about forty years.

The Koktokay mine has produced a number of minerals including allanite-Ce, albite, amazonite, apatite, aquamarine, ashanite, bismuthinite, bismutomicrolite, chrysoberyl, epidote, ertixiite, fluorite, garnet, hiddenite, holmquistite, ilmenite, ishikawaite, manganotantalite, manganocolumbite, molybdenite, muscovite, phenakite, pollucite, polylithionite, pyrite, qingheiite, samarskite, scheelite, sphalerite, spodumene, staurolite, topaz, tourmaline, trilithionite/polylithionite, uraninite and zircon. Koktokay is the type locality for three minerals: ashanite, ertixiite and qingheiite. The mine also produced the first known Chinese occurrences of the rare minerals chabazite-Na and ishikawaite.

Because of the strategic importance of this state-run mine, the Koktokay area was under military control until the 1980s, and the town of Koktokay could not be found on maps published prior to the late 1990s. Until that time, any gem-quality minerals recovered from the mine were handled as ore or as waste. When I took my first trip to Koktokay in 1996, many bluish, transparent beryl fragments were visibly scattered across the open ore piles. In general, the miners were unaware of the gemstone value of these fragments. In 2000, the Koktokay mine closed and nearly one-third of the population of this small mountain town left the area. The abandoned open pit is currently filled with water.

Today, in the city of Altai and the towns of Fuhai, Fuyun and Qinghe, jewelers process and trade the aquamarine, gemstones and raw materials found in the area mines. More than 70 percent of the China's aquamarine jewelry is manufactured in the Altai area, and more than 50 percent of China's gemstone aquamarine originates in the Altai region; still, over several visits to the area, the author has never seen a single top collector's specimen in Altai’s markets.

South of the Clouds

Yunnan Province is another important beryl-producing region in China. Yunnan, which means "South of the Clouds" is China's most southwest province, bordering the countries of Vietnam, Laos and Burma. Guizhou Province and Guangxi Zhuang Autonomous Region are its neighbors to the east, Chongqing and Sichuan to the north, and the Tibet Autonomous Region borders Yunnan in

China Blue

This 9 cm high, etched blue beryl is from a mine in the Gaoligongshan Mountains, Yunnan.

Kevin Dixon collection and photo

the northwest. Yunnan encompasses 394,000 square kilometers (152,084 square miles) and has a population of more than 42 million people.

The province enjoys a diverse topography that ranges from high mountain ranges to tropical rainforests. Yunnan is home to the greatest variety of plant species (more than 18,000 identified) in China as well as to an incredible array of animals, including the Asian elephant and the protected Yunnan golden monkey. High mountains, some of which are snow-capped year round, characterize the western and southeastern portions of Yunnan.

Located on the outskirts of the capital city of Kunming, Stone Forest is a natural wonder in which multiple, massive limestone pillars have formed a gigantic forest of stone, showcasing the spectacular karst geography. Yunnan's natural features such as deep virgin forests, lakes and hot springs, precipitous valleys, rivers and waterfalls attract tourists from around the world.

There are 25 ethnic minorities in Yunnan. This multiplicity of cultures has endowed the province with a rich heritage of different religions and colorful customs.

Yunnan's climate varies widely with latitude and elevation, from subtropical in the southern and flat areas to cold tundra in northern and western highland areas.

The Gaoligongshan and Ailaoshan mountain belts in the western portion of Yunnan are the province's primary gemstone localities. These two uplift belts were formed during the Mesozoic and Tertiary as a result of the continental collisions between the Eurasian and Indian plates. Gaoligongshan Mountain is the easternmost extension of the Himalayan range. Gemstone-rich pegmatites and skarns are widely distributed throughout these two mountain belts. Only a limited amount of geologic work has been done in these areas and reliable detailed data are unavailable. Gem mining activities are limited to a smattering of exposures in the area.

This summer view of the Tibetan village of Huya in Sichuan, home to most of the beryl, scheelite and cassiterite miners, was taken by the author.

In the **Ailaoshan Mountains**, gem minerals generally formed within vugs found in the pegmatite veins and granites. Beryl from this area is sometimes found in association with topaz. Crystals tend to be very well developed and make excellent specimens. Unfortunately, miners in Yunnan, as in much of China, tend to focus their efforts on recovering gem rough, rarely saving crystal specimens.

Aquamarine and other gemstones including topaz, apatite, zircon, tourmaline and cassiterite are commonly found in the Ailaoshan area. These species mainly occur in the areas around **Yuanyang**, **Honghe**, **Yuanjiang** and **Jingping**. In Yuanyang, aquamarine crystals generally range in size from 0.2 to 4 centimeters wide and 0.5 to 20 centimeters long. These exhibit intensive blue color and have sometimes developed with gas and fluid inclusions. Occasionally, yellow and pink beryl is also found in the area.

Beryl from the **Gaoligongshan Mountains** is generally associated with tourmaline, pollucite, epidote and others. The main known beryl localities are in the vicinities of **Ruili**, **Gongshan**, **Longling** and **Fengqing**. During the 1960s, a few state-run mines in the area were being worked for beryllium, niobium-tantalum and other ores. Most of the gemstones that were recovered during those mining operations were either used as ore or were relegated to the dumps. Formal mining operations were terminated in the 1970s. Today, local farmers casually work the abandoned mines and pits.

Naturally etched (skeletal) aquamarine crystals that are seen on the market generally come from the Gaoligongshan localities, which produce what is currently considered China's best aquamarine specimens. Crystals from Gaoligongshan range from several centimeters to 50 centimeters in length. The largest crystals, however, tend to be heavily fractured.

China's emerald is generally produced in the **Maguan-Malipo** area of southeastern Yunnan, although it has also been found in **Heilongjiang** and **Qinghai**. The most important locality for emerald is **Mengdong**, about 70 kilometers southwest of Malipo.

Mengdong's emerald deposits mainly formed

in the metamorphic zone between the lower Palaeozoic sedimentary rocks and intrusion bodies. The emerald is a product of hydrothermal mineralization. Mengdong emerald is found in association with scheelite, pyrite and fluorite and is generally "frozen" in quartz veins that run through the metamorphic rocks. Crystals can be very large, sometimes reaching thicknesses of up to 5 centimeters and lengths of up to 30. They are deep green in color but tend to be highly fractured and are not suitable for use in jewelry.

Four Rivers

There is a great deal more than mineral specimens to entice collectors to Sichuan Province, one of the largest provinces in the nation. Home of the Giant Panda, this province covers 485,000 square kilometers (187,000 miles) of southwest China and boasts the contry's largest population (87 million people). Sichuan means "four rivers."

Sichuan is bordered by the Tibetan Plateau in the west, the Yunnan-Guizhou Plateau in the south and east, and Qinling Mountain in the north. Sichuan is thus a huge basin. The Yangtze River cuts through the province creating spectacular scenery, including the Three Gorges. In fact, Sichuan is home to four UNESCO world heritage sites: the Jiuzhaigou Valley Scenic and Historic Interest Area, the Huanglong Scenic and Historic Interest Area, the Mount Emei Scenic Area including the Leshan Giant Buddha Scenic Area, and Mount Qingcheng and the Dujiangyan Irrigation System. The eastern part of the Sichuan is dominated by a subtropical and humid climate, while the western highlands see higher rainfalls, intense sunlight and lower temperatures. Sichuan's annual rainfall averages 100 centimeters. The average annual temperature is 16.5 degrees Celsius.

Sichuan is populated by the Hans as well as some 15 other minor ethnic groups. These minorities are predominately Yi, Hui, Qiang, Tibetan and Miao. The province's burning hot cuisine provides some for the spiciest dishes on the planet, seasoned with local chilies, wild pepper and garlic.

Although aquamarine has reportedly been found in many Sichuan localities (including **Wenchuan**, **Danba**, **Kangding**, **Ya'an** and **Jiulong**), **Xuebaoding Mountain** near Pingwu County is the most important locality in the province for beryl. Xuebaoding Mountain is situated in the northwestern portion of Sichuan and is a part of a transitional belt between the Sichuan basin and Tibetan plateau. Peaking at an altitude of 5,588 meters, Xuebaoding Mountain is famous for producing world-class red to orange scheelite, huge, lustrous cassiterite and pale, short hexagonal beryl specimens.

Sharp!
This 5.3 cm high, gemmy, face-rich beryl crystal on matrix is from Xuebaoding Mountain near Pingwu County in Sichuan.
Collection Marcus Budil; photo Jeff Scovil

Xuebaoding Mountain encompasses a large area of ridges over which numerous quarries for scheelite, cassiterite and beryl are scattered. Though these mines and prospects may have been locally given individual names, they are generally working the same massive deposit. That said, over the last two years, some scheelite and cassiterite specimens have been recovered from an area *near* Xuebaoding Mountain. These are likely from a different deposit as the specimens exhibit different characteristics and mineral associations.

It takes roughly 10 hours to drive the 320 kilometers from Chengdu, the capital of Sichuan, to the Xuebaoding locality, which is not far from the predominantly Tibetan village of Huya.

Xuebaoding is a tungsten-cassiterite-mica-beryl deposit that mainly formed by hydrothermal

Qinghai
Sichuan
Xizang Zizhiqu
Yunnan
Tanglag
Sangruma
Darlag
Maqu
Jigzhi
Zoigê
Chinqên
Xiwu
Cumda
Sêrxü
Baima
Aba (Ngawa)
Hongyuan
Songpan
Qagca
Darcang
Nyidoi
Dainkog
Dainkognubma
Jumanggoin
Gosu
Sêrtar
Zamtang
Zogqên
Catung
Sinda
Maniganggo
Heishui
Barkam
Dêgê
Dainma
Garzê
Miyaluo
Jomda
Qu'nyido
Jinchuan
(Quqên)
Luhuo
(Zhaggo)
Baiyü
Gonjo
Xinlong
(Nyagrong)
Dawu
Xiaojin
(Zainlha)
Bawê
Garyi
Danba
(Rongzhag)
Mt. Qigcheng
& Dujiangyan
Zhag'yab
Kajing
(Nyaquka)
Kangding
(Dardo)
(Jagsamka)
Batang
Litang
Luding
Ya'an
Zogang
Markan
Chubalung
Bokog
Mt. Emei
Chagyoi
Jiju
Sudmo
Daocheng
(Dabba)
Jiulong
(Gyaisi)
Ganluo
Xiangcheng
(Qagchêng)
Ragwa
Püyog
Wengshui
Bawolung
Mabian Yizu Zizhix
Dêrong
Dêqên
Mianning
Yuexi
Xide
Meigu
Nixi
Shangrila
Muli Zangzu Zizhixian
Xichang
Zhaojue
Gongshan Drungzu
Nuzu Zizhixian
Luguhu
Butuo
Yanyuan
Baijixun
Dechang
Puge
Tuguancun
Weixi Lisuzu
Zizhixian
Ninglang Yizu Zizhixian
Ningnan
Fugong
Yulong Naxizu
Zizhixian
Miyi
Qiaojia
Geliping
Jianchuan
Heqing
Yongshen
Yanbian
Huili
Huidong
Yingpan
Lanping
Baizu Primizu Zizhian
Huaping
Panzhiua
Wanbi
Eryuan
Pingchuan
Santai
Lushui
Yunlong

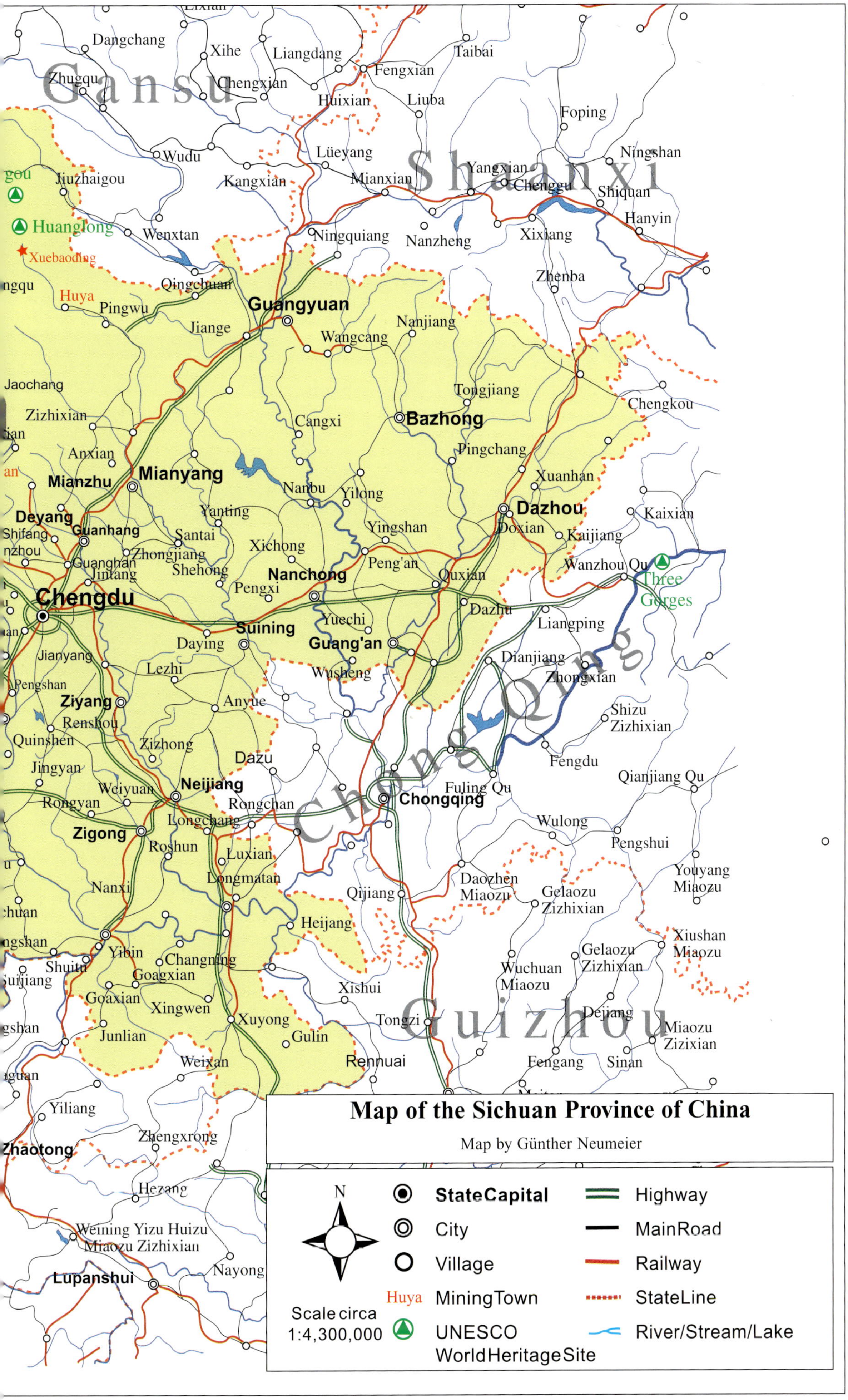
Gansu
Shaanxi
Chong Qing
Guizhou
Dangchang
Xihe
Liangdang
Fengxian
Taibai
Zhugqu
Chengxian
Huixian
Liuba
Foping
Wudu
Lüeyang
Ningshan
Jiuzhaigou
Kangxian
Mianxian
Yangxian
Chenggu
Shiquan
Huanglong
Wenxtan
Ningquiang
Nanzheng
Xixiang
Hanyin
Xuebaoding
Zhenba
Huya
Pingwu
Qingchuan
Guangyuan
Jiange
Wangcang
Nanjiang
Jaochang
Tongjiang
Chengkou
Zizhixian
Cangxi
Bazhong
Anxian
Pingchang
Mianzhu
Mianyang
Nanbu
Yilong
Xuanhan
Deyang
Yanting
Dazhou
Kaixian
Shifang
Guanhang
Santai
Yingshan
Doxian
Kaijiang
Zhongjiang
Xichong
Guanghan
Shehong
Peng'an
Wanzhou Qu
Three Gorges
Jintang
Nanchong
Quxian
Chengdu
Pengxi
Dazhu
Yuechi
Liangping
Suining
Daying
Guang'an
Jianyang
Dianjiang
Lezhi
Wusheng
Zhongxian
Pengshan
Ziyang
Anyue
Shizu Zizhixian
Renshou
Quinshen
Zizhong
Fengdu
Dazu
Jingyan
Qianjiang Qu
Weiyuan
Neijiang
Fuling Qu
Rongyan
Rongchan
Chongqing
Longchang
Wulong
Zigong
Pengshui
Roshun
Luxian
Youyang Miaozu
Nanxi
Longmatan
Daozhen Miaozu
Qijiang
Gelaozu Zizhixian
Heijang
Xiushan Miaozu
Yibin
Changning
Gelaozu Zizhixian
Shuitu
Goagxian
Wuchuan Miaozu
Xishui
Goaxian
Xingwen
Dejiang
Xuyong
Tongzi
Miaozu Zizixian
Junlian
Gulin
Rennuai
Fengang
Sinan
Weixan
Yiliang
Zhengxrong
Zhaotong
Hezang
Weining Yizu Huizu Miaozu Zizhixian
Nayong
Lupanshui
Map of the Sichuan Province of China
Map by Günther Neumeier
N
StateCapital
City
Village
Huya MiningTown
UNESCO WorldHeritageSite
Highway
MainRoad
Railway
StateLine
River/Stream/Lake
Scale circa 1:4,300,000

This sharp, color-zoned beryl on albite was part of a lot of about 20 specimens that came out of a Pingwu County, Sichuan pegmatite in 2000. The main crystal is 2 cm wide. Collection and photo Kevin Dixon

processes. The formation and distribution of the deposits are controlled by the framework of fracture systems, primarily occurring toward the top of the mountain. The government began to mine the area in the 1980s for tungsten and tin ores, and in the 1990s, Xuebaoding became famous for scheelite and cassiterite.

Beryl from the area generally forms as transparent, colorless, short hexagonal crystals. The typically tabular hexagon habit is very uncommon for very light blue or colorless beryl. These crystals have thus captured the attention of international academics. Single crystals range in size from less than one centimeter to more 20 centimeters across. Producing almost exclusively blue or colorless beryl, the locality surprised miners a few years ago with one 5-centimeter wide morganite crystal. Until the late-1990s, the beryl at Xuebaoding was only mined for beryllium ore, and to date only a very small number of colorless crystals have been sold for jewelry.

South of the Lake

Located in the south central part of China, Hunan province has long been known for its natural beauty. With abundant natural resources, Hunan covers 210,000 square kilometers (81,000 square miles). It is adjacent to Jiangxi Province in the east, Chongqing Municipality and Guizhou Province in the west, Guangdong Province and Guangxi Autonomous Region in the south, and Hubei Province in the north. Hunan is characterized by its many mountains and hills, and by five major rivers that flow into the Yangtze River in the north. In 1982, China's first national park, Wulingyuan (Zhangjiajie), was established in a particularly scenic portion of northwest Hunan. The park is known for its quartz-sandstone forests, sculpted by natural weathering and vertical erosion. Hunan has a humid, subtropical climate, with monsoons in April, May, and June. July and August are uncomfortably hot and humid. For thousands of years, Hunan has been a major production center of minerals, rice, tea and oranges.

Beryl variety Goshenite
3 cm; Xuebaoding mountain, Pingwu, Sichuan, China
Fine Minerals International; Jeff Scovil photo

Jeff Scovil captured this view of a pegmatite outcropping at Chenjiawang, Hunan on an April 2004 trip to China.

Hunan has a total population of 64 million, consisting of mainly Han Chinese with some minorities such as the Miao, Dong and Yao peoples. Many famous politicians including Mao Zedong, the founder of the People's Republic of China, came from Hunan Province.

Because of its abundant mineral resources and active mining activities, Hunan has become the center of the Chinese mineral specimen trade. Most of China's top mineral specimens come from Hunan's mines and most Chinese mineral dealers are native to Hunan.

Granitic pegmatite-type beryl is widely found in Pingjiang County in the northeastern portion of Hunan Province. Pegmatite minerals are distributed mainly in the Nanjiangzhen and Meixianzhen areas of northern Pingjiang. There are a number of mines working in the area. These include the **Mugualong feldspar mine** in Hupai, 44 km from the town of Pingjiang and the **Nanjiangjiao beryl mine** near the town of Nanjiangzhen.

From the early 1960s until the late 1990s, all of the area mines were state or province-run beryl and feldspar mines and were exploited for industrial minerals. Gem-quality beryl was often found in these mines, which sometimes produced chatoyant crystals. This area may be the one of the first localities in China known to produce aquamarine. Since the 1970s, a handful of dealers and jewelers have been buying stones from local miners.

Mines in Pingjiang County have produced some notably large beryl crystals. The largest recorded weighed 2,340 kilograms, and was found in the Mugualong mine in the 1980s. The specimen is now exhibited in China's Geological Museum in Beijing.

In the late 1990s, most of the state and province-run mines in Pingjiang's pegmatite areas were closed either because of a slowdown in the economy or because the ore deposits were exhausted. Gemstones are uncommon enough in these mines that it is not economically feasible to reopen them solely to mine for gems or for specimens. As a result, many ex-miners and part-time dealers have moved to Guilin and Guangxi to become professional stone and mineral dealers. On a visit to the area in April 2004, I observed that many small mines and open pits were being operated by local private concerns or by townships. The mines were being worked primarily for industrial feldspar, and some large, blue opaque beryl crystals were being produced.

Upper Left: *This 9.2 cm high specimen from the author's collection is an example of the beryl produced in Pingjiang, Hunan.*

Upper Right: *A dealer in Chenjiawang, Hunan shows off a beryl specimen on matrix.*

Both photos Jeff Scovil

Other Localities

In addition to China's most significant beryl-producing provinces, deposits in Hubei, Heilongjiang, Jiangxi and other provinces have also been known to now and again yield colored gem beryl. Most of these localities are undeveloped and there is very limited information available about them. No doubt, many of these areas have the potential to be significant future sources for gem beryl.

The **Tongchang pegmatite** in Hubei Province, for example, could be an important future source for beryl. The Hubei Provincial Museum recently acquired two high-quality aquamarine crystals from Tongchang. One of the two is a 10 x 5 centimeter, deep blue crystal that is free of inclusions.

Xushui, **Xingzi**, **Yifeng** and **Gangxian** counties in Jiangxi province have been producing small quantities of low-quality beryl, but who knows what the future will hold?

Geological exploration in the **Heihe**, **Linkou**, **Luobei**, **Boli** and **Jiamusi** areas of Heilongjiang Province has uncovered a number of deposits bearing blue, gem beryl. Some of these deposits also host green beryl. All of the deposits are pegmatitic and most are as yet undeveloped.

Central Inner Mongolia, including the areas of **Alashan**, **Wuteila** and **Huade**, is an area rich in pegmatites. Beryl occurs there in a wide range of quality and with a variety of associations. The region has produced large (5 x 30 centimeter) crystals of gem aquamarine. Because most of the Inner Mongolian deposits are located in remote and economically undeveloped areas, gemstone mining in the region is still in its infancy.

Gem Beryl in Australia

Dermot Henry, Senior Collection Manager at Australia's Museum Victoria, with a survey of Australia's gem beryl deposits

Gem-quality beryl has been unearthed from pegmatites throughout Australia. While many green beryls are called "emeralds," most owe their color to the presence of vanadium or copper rather than chromium (Sutherland, 1991). A few "true" emeralds have been mined from pegmatites near Emmaville and Torrington in the New England area of New South Wales. Emerald has also been mined from schists adjacent to pegmatites at Poona and west of Menzies in Western Australia. Beryl has also been mined as an industrial mineral on several locations in Australia.

New South Wales

Many fine beryl specimens have been obtained in the **Emmaville** and **Torrington** districts of the New England region of northern New South Wales, with over 35 beryl-emerald occurrences identified in the district (Henley & Brown, 2000). Green beryl and the gem varieties emerald and heliodor were discovered during mining for tin and bismuth-tungsten in the 1890s. Curran (1897) reported that "Some magnificent crystals have been found, some of the best of which have, to the writer's knowledge, been acquired by foreign museums."

A 5 cm long beryl crystal from Torrington, New South Wales
Museum Victoria specimen No. M23109; Photo John Haupt

The Emerald Proprietary Company was formed in 1890 to exploit the discoveries of emerald in the De Milhous reef near Emmaville. A parcel of 2,000 carats of rough and 10 carats of cut and polished emeralds was sent to London for valuation (Webb & Sutherland, 2002). This valuation encouraged the development of the mine, which operated between 1890 and 1909. Sporadic mining occurred until 1963, when the mine was reopened and a substantial amount of beryl and emerald was produced until closure in 1964.

At **Fielders Hill**, 9 kilometers northwest of Torrington, emerald occurs as strongly defined zones in crystals of colorless beryl. Parts of these crystals may be cut perpendicular to the *c*-axis to produce a standard, emerald-cut gemstone.

Some fine examples of colorless to bright green beryl were collected from **Heffernan's wolfram mine**, 5 kilometers west of Torrington. Samples of gem quality yellow beryl were also collected from the Torrington area (Webb & Sutherland, 2002).

Beryl has been mined from pegmatites in the Broken Hill-Euriowie district of New South Wales. The largest deposit was the **Triple Chance mine**, famous for the occurrence of large, hexagonal, prismatic crystals over 30 centimeters long. The mine produced more than 339 tons of beryl during its operation (Olliver & Stevenson, 1984).

Northern Territory

There are many beryl-bearing pegmatites in the Northern Territory although there are few records of gem-quality specimens being produced. Pegmatites mined for mica in the Harts Range (e.g., **Delma mica mine**) and in the Strangway Ranges (e.g., **Annamurra Creek** area) have produced beryl crystals over 30 centimeters in length.

Queensland

Beryl has been recorded from a number of localities in Queensland, the most noteworthy occurrence being in the **O'Briens Creek** area near Mt. Surprise, 200 kilometers southwest of Cairns in north Queensland. Established in 1995 by the Department of Mines and Energy, the O'Briens Creek Fossicking Area is a popular collecting site for large topaz and beryl crystals. Blue beryl crystals up to 10 centimeters long have been collected from pegmatite-derived alluvial gravels that are up to 2 meters deep.

South Australia

In the South Australian portion of the Olary province some 136 tons of beryl were produced from pegmatite deposits mainly as a byproduct of feldspar mining. No gem-quality specimens have been recorded.

Tasmania

Beryl associated with tin-bearing granites occurs in several areas of Tasmania. Attractive pale blue to green crystal groups up to several centimeters long occur at the famous **Mt. Bischoff tin mine** on the west coast of Tasmania. In the northeastern portion of the state, in the Moina region, pale green beryl crystals up to a few centimeters in length have been recorded. Near Ben Lomond, large well-formed crystal groups with individual crystals up to 25 centimeters long have been noted (Petterd, 1910).

A 8.5 cm long beryl crystal from O'Briens Creek, Mt Surprise, Queensland

Australian Museum collection; photo Frank Coffa

Victoria

There are only a few minor occurrences of beryl in Victoria, however, a one-of-a-kind find consisting of yellow beryl crystals was discovered in a granite pegmatite in a gravel pit near Mt Bolton and Ballarat (Birch & Henry, 1997). The transparent beryl crystals, up to 6 centimeters long, occurred in a cavity lined with smoky quartz crystals. The beryl crystals are striated along the *c*-axis and the terminations have been removed by natural etching.

Western Australia

The Poona region, approximately 60 kilometers northwest of Cue, is the source of most of Western Australia's commercial emerald production. The field was discovered by tin prospectors in 1909. Beryl and emerald occur in both Precambrian biotite schist and quartz-beryl pegmatites. The best emeralds are found in the biotite schist adjacent to the pegmatite (Graindorge, 1974). The principal mine in the region, the **Aga Khan Deep mine**, has been intermittently worked both as an open cut and more recently (1977-1981) as an underground operation (Palmer, 1990). Smaller emerald deposits have been worked in the district with the **Emerald Pool** and **Poona East Emerald mine** being the most productive.

Emerald has also been commercially produced at the **Menzies mine** located on the Riverina Station, about 50 kilometers west of Menzies in the Yilgarn Craton. Beryl occurs in two pegmatite veins that cut lenticular chlorite schist. The pegmatite contains abundant pale green opaque beryl crystals while "true" emerald crystals, up to 2 centimeters long, occur along the contact zone between the chlorite schist and the pegmatite.

In the Pilbara area, beryl has been reported from a number of localities (Jacobson et al. 2004) with the **Curlew emerald deposit** producing some of the best specimens for collectors. The mine is located 19 kilometers northwest of Hillside Station near Marble Bar. Beryl occurs in pegmatite veins and in adjacent biotite schists. The mine produced several thousand carats of rough emerald and beryl with the report of one large gem-quality emerald weighing 600 carats (Jacobson et al. 2004).

The **Tabba Tabba Main pegmatite**, which has been mined for tantalum, has produced some unusual white, lustrous, doubly terminated beryl crystals. Typically the crystals are up to 5 centimeters across with the largest crystal recorded measuring 28 centimeters in diameter and 18 centimeters along the *c*-axis (Jacobson et al. 2004).

While gem-quality pink beryl has not been recorded from Australia, some pale pink beryl crystals up to 4 centimeters across have been collected from the **Londonderry feldspar quarry** within the Coolgardie pegmatite field (Jacobson, 1997). Occasional aquamarine has also been found in the district.

Beryl from Southern Africa

Bruce Cairncross chairs the Geology Department at University of Johannesburg in Auckland Park, South Africa. He authored the **Field Guide to Rocks and Minerals of Southern Africa,** *published in November 2004*

Beryl occurs in most southern African countries. The most noteworthy localities are found in Mozambique, Namibia, Zambia and Zimbabwe. Superb specimens of aquamarine have come from pegmatites in these countries, and excellent quality emerald has been known for decades from Zambia and Zimbabwe. The following is a survey of some of southern Africa's major deposits of beryl and its color varieties: aquamarine, emerald, heliodor, morganite and goshenite.

Angola

Beryl deposits are known from lepidolite-spodumene-bearing pegmatites located approximately 100 kilometers east of the capital Luanda. Economic deposits of beryl from these pegmatites were exploited during the early 1970s east of the town of **Narnibe**, the capital of the Narnibe province. Pegmatites in proterozoic gneiss and schists exist in two varieties there: (i) microcline-quartz with muscovite and tourmaline and (ii) lepidolite-spodumene with tantalite-columbite, the latter hosting the beryl mineralization. Three hundred eighty tons of beryl were produced between 1969 and 1973.

Democratic Republic of the Congo

Mineralized lithium-tin-tungsten bearing pegmatites are relatively common west of Lake Kivu. The main beryl-rich pegmatites are found at **Kobokobo** near Kamituga, about 180 kilometers southwest of the lake. There beryl is most commonly associated with schorl.

Kenya

There are only scattered occurrences of beryl in Kenya (Keller, 1992). Aquamarine has been sporadically mined at the **Sebit mine** south of Lake Turkana. The **Embu** district produces gem blue beryl together with topaz and amazonite. Other beryl localities are scattered along the **Thura River**, southern **Machakos** district and in the **Kianbere** area.

Malawi

In northern Malawi's **Mzimba** district, swarms of pegmatites and quartz veins occur in gneiss and schist. Some of these host beryl, including some low-quality emerald, which is associated with green elbaite, schorl, rose quartz and amethyst. This pegmatite belt extends into Zambia in the **Lundazi** area, which is known to have deposits of aquamarine and emerald.

Mozambique

The **Alto Ligonha** pegmatites in Mozambique (Hutchinson & Claus, 1956) have yielded many fine gemstones, including beryl and its color varieties morganite (rose-red to salmon-pink), aquamarine, goshenite and emerald, the latter hosted mainly in schist. An unusual black variety of beryl is also found in the region. Bettencourt, Dias & Wilson (2000) list 27 pegmatite groups in the area, and beryl is found in 20 of these. The pegmatite belt occurs between Ribaue and Alto Molócue. Individual crystals and masses can weigh up to 900 kilograms, with crystals often more than 1 meter in diameter.

The Muiane pegmatite in Alto Lighona, Mozambique in 1993. The opencast area is about 400 meters across. Benches have exposed the highly kaolinized deposit. Paul Balayer photo

The **Muiane** pegmatite is one of the largest in southern Africa. It has infrequently produced excellent quality beryl, aquamarine and morganite. Black beryl from this mine is colored by manganese inclusions. Nearby pegmatites have yielded goshenite, aquamarine and fine peach-colored morganite. The **Maria II mine** produced emeralds, and the **Niame emerald mine** near Gilé produced emerald crystals up to 25 centimeters

long (Bancroft, 1984). The Muiane mine was recently re-opened and still holds great potential for future production (personal communication, Paul Balayer, 2004). At the **Namivo mine** northeast of Gilé, a large mass of intergrown morganite and aquamarine yielded 200 tons of beryl (von Knorring, 1970).

Namibia

Beryl tends to be concentrated in pegmatites in a broad region extending from Brandberg West-Uis in the north, to Sandamap-Karibib-Usakos in the south, as well as in Tantalite Valley in the south of the country. Over 50 Namibian pegmatites have been worked for beryl.

Namibia has some of the finest deposits of aquamarine in the world. These come particularly from miarolitic pegmatites located in the Karibib-Usakos district. For example, superb blue crystals were collected from **Davib West 62** and **Erongorus 166** in the **Erongo Mountain**, where crystals up to 20 centimeters long and 5 centimeters in diameter are found associated with black schorl tourmaline, white feldspar, quartz and green fluorite. The most sought-after aquamarine is color zoned from pale blue at the base to darker blue toward the termination of the crystal. This color trend is coupled with an increase in the clarity in the crystal, from opaque at the base, to transparent in the upper sections.

From **Klein Spitzkoppe**, located west of Erongo, beautiful aquamarine associated with smoky quartz, topaz, phenakite and fluorite has been found for decades. The **Rössing** pegmatites, 5 kilometers north of Rössing station, and the farm **Donkerhoek 91** in the Karibib district are noteworthy localities for beryl and aquamarine. Pegmatites in the **Namib Naukluft Park**, for example, the **Riet 30 farm** and

Alto Ligonha

Left: *An 8.1 cm high aquamarine crystal from Alto Ligonha in Mozambique; Gene Meieran collection*

Right: *This 10.2 cm wide morganite crystal was being offered for sale by Jendon Minerals in 2000.*

Both photos Jeff Scovil

Klein Spitzkoppe in Central Namibia

The pegmatites in this granite have produced outstanding aquamarine and topaz crystals.

Bruce Cairncross photo

Right: *Namibia's Erongo Mountain is known for producing classic pegmatite associations such as this wonderful 6.5 cm wide beryl and schorl grouping found at Bersig 274.*

Collection Harold Muench; photo Jeff Scovil

Lower Right: *Klein Spitzkoppe in Namibia is the source of this 3.7 cm high beryl now in the Museum Afrika collection. Photo by the author*

Roadside Collecting

In March and December of 1999, I had the opportunity to visit the **Klein Spitzkoppe** in Namibia. Collecting minerals in the Klein Spitzkoppe granite is not easy, and it's better to travel there with a "silver pick" in the hope of buying specimens.

One drives on the main road from Windhoek to Walvis Bay, and the impressive Spitzkoppe mountains can be seen to the right after you pass the towns of Karibib and Usakos. A gravel road to Henties Bay turns north off the main road and leads to the locality. Local Ovambo diggers have informal shacks in the area and they dig for crystals in the outcrop. Their families sit alongside the road with makeshift tables, where one can stop and buy specimens.

Some of the best examples of Klein Spitzkoppe minerals in my collection were bought at these roadside "stops" where women and children offer specimens. Although topaz, feldspar and quartz dominate the offerings, some beryl and aquamarine may also be found. The prices (in 1999) were very reasonable compared to similar quality specimens sold by dealers, and it wasn't necessary to haggle for cheaper prices because the community there is impoverished and desperately in need of money. Nowadays, however, one needs to be careful when buying at the roadside because specimens from other localities such as the nearby Erongo Mountain and even the Brandberg to the north are mixed with the Klein Spitzkoppe material.

Wilsonfontein 110, about 120 km east of Swakopmund also produce aquamarine. In the Omaruru district, very good quality aquamarine came from a pegmatite on the **Kawab 117** farm. Some beryl crystals up to 3 meters long and 1.5 meters in diameter have come from the **Etiro** pegmatite.

The term "heliodor" was originally coined to refer to beryl discovered in 1910 in Namibia in the **Hoffnungsstrahl** pegmatite, situated close to Rössing Siding. Some of the faceted material from this deposit was given as a gift to Empress Victoria of Germany. Attractive heliodor specimens have been collected in the Erongo pegmatites and at Klein Spitzkoppe. The latter locality has produced crystals up to 12 centimeters long and 5 centimeters in diameter that are associated with fluorite. Heliodor is also found in the **Okahandja** townlands.

Pink beryl crystals come from the **Rubikon** and **Helikon** pegmatites in the **Karibib** district

The Gravelotte Emerald mine in South Africa is no longer operating, but in its heyday it produced some high-quality gemstones.

Bruce Cairncross photo

and from **Neu Schwaben**. Morganite is also reported from the **Etiro 50 farm** in the Karibib district. A small emerald deposit is located on the **Neuhof 100 farm** in the Maltahöhe district.

Colorless beryl has been collected from a pegmatite in the **Omaruru** district and from various other pegmatites on Erongo Mountain.

Rwanda

This country has lithium-bearing and cassiterite-rich pegmatites, but beryl appears to be relatively rare (von Knorring, 1970).

South Africa

Beryl is one of the most common minerals found in the pegmatites of the Northern Cape Province. At the **Middel Post** pegmatite in the Keimoes district, beryl crystals from the different parts of the pegmatite are different colors, ranging from blue-yellowish green to smoky gray to milky white to deep rose and transparent blue. In lithium-rich pegmatites, the beryl is usually white to pink. It commonly occurs as subhedral to euhedral hexagonal crystals, from very small to more than a meter long. In some pegmatites, instead of well-formed beryl crystals, large masses weighing over 10 tons are found. Some of the more complex pegmatites contain irregular masses of aquamarine. The **Angelierspan** pegmatite was the most productive beryl source in the region. In the 1950s, a 60-ton beryl crystal was exposed there, and 2-meter long crystals were common.

In the Limpopo and Mpumalanga Provinces, beryl is a common pegmatite mineral, particularly in several deposits in the **Musina** and **Soutpansberg** districts. Beryl occurs in the Archaean rocks in the **Letaba** district. About 20 kilometers from Mooketsi Station, on the **Welgevonden 886 LS farm**, beryl is found in a small pegmatite in crystal clusters up to half meter wide with individual crystals 8 centimeters in diameter.

Several pegmatites in the **Polokwane** district are beryl-bearing, for example the farm **Verbrandhoek 983 LS**, 65 kilometers east of the town; at **Kalkfontein 615 LS**, 25 kilometers north of the town; and at several smaller pegmatites southwest of town. At Kalkfontein, two pegmatites contain beryl crystals that can reach 15 centimeters in diameter. These are associated with quartz, albite and spodumene.

Beryl was mined at the **Witkop pegmatite** in the **Piet Retief** district, Mpumalanga. Aquamarine is found in **Namaqualand** in the Northern Cape Province near Jakkalswater. Peach-pink to rose-pink morganite has been found in pegmatites at **Steyns Puts** (Namaqualand) and near **Leydsdorp** in the Limpopo Province, and heliodor has been reported from the farm **Kitchener 504 MS** located between Musina and Polokwane.

Emerald has been mined at **Gravelotte**, at the **Cobra Emerald mine** in the Murchison greenstone belt. The emerald is found in pegmatites that crosscut biotite, chlorite and actinolite schists. The best crystals are usually found in the biotite schist close to the contact with pegmatites. Crystals are mainly small but some reach 5 centimeters. Stones up to 10 carats have been faceted, though most are opaque and only of specimen value. Production at the emerald mines in Gravelotte was most significant in the early 20th century. During an 8-year period in the 1930s,

The "Night Shift" at Gravelotte

Traveling to the Gravelotte emerald mines is a pleasant trip. It's about a 4-hour drive from Johannesburg on paved roads, and once you descend the escarpment, the road passes though typical, beautiful thornveld scenery and rolling hills.

When we visited the **Cobra Emerald mine** in the early 1990s, it was no longer being worked, and we had access to the entire area. The old workings are extensive and it took a while to find any indication of emerald in the sprawling schist outcrop that dominates the open pit. Soon, however, I saw signs of "unofficial mining"—locals who trespass at night to scratch around for the odd gemstone—and decided to look closer at their diggings.

They were near a series of quartz veins, and sure enough, close to the contact between the quartz and the host rock, tiny specks of green emerald could be seen. It was fairly easy to dig these out of the enclosing soft schist but the crystals cleave very easily; so it proved more worthwhile to remove the emeralds *in situ* which makes for better specimens anyway. We made no major discoveries that day but it was an enjoyable time spent in South Africa's famous old emerald deposit.

From time to time, some specimen-grade emerald still appears on the market in South Africa. These are not gem quality but provide evidence that the night shift is still operating!

664,585 carats of emerald were mined. Emerald occurred in association with feldspar, molybdenite, pyrite, blue apatite, schorl and quartz. Emerald has also been mined at the **Phalmartin Emerald mine**, and some emerald was found in biotite schist at **Uitvalskop** near Schweizer-Reneke.

Swaziland

Beryl is found in granitic pegmatites in the foothills of the **Sinceni Mountains**. The largest beryl crystal found here weighed 106 kilograms, and crystals weighing 5 to 15 kilograms are common. Beryl is associated with quartz, muscovite, albite and microcline, as well as lesser amounts of topaz, cassiterite, magnetite, tantalite, schorl, monazite and zircon.

Tanzania

Emerald comes from the **Lake Manyara** region in the Mbula district, Arusha region, where emerald and alexandrite crystals occur in mica schist in aggregates along contact zones between biotite-actinolite schist and pegmatites.

Uganda

In southwest Uganda, beryl pegmatites have been exploited on a major scale. In 1961, for example, 1,000 tons of beryl were produced (von Knorring, 1970). Pegmatites are commonly kaolinized, and beryl is associated with other economic species including cassiterite, ferberite, spodumene, lepidolite and lithium-phosphate amblygonite (von Knorring, 1970). Euhedral beryl crystals vary from small to more than 1 meter long. Irregular masses are also encountered, such as one weighing approximately 50 tons found at **Kabira**.

Zambia

Zambia is relatively well known for emerald and aquamarine. Emerald was discovered in the Miku area during the late 1920s but was not commercially exploited until 1967. Some notable emerald deposits that have been and are still being exploited are in the **Ndola** rural district, namely the **Kagem**, **Kamakanga**, **Pirala** and **Fwaya-Fwaya** deposits. The **Kafubu** emerald deposit contains gemstones concentrated along the contact margins between talc-chlorite-tremolite schists and quartz-tremolite pegmatites, the latter intruding into the host schist. The source of the chromium in the emerald was

Right: *Italian dealer Gaetano Lacagnina brought this impressive, gemmy, 15.3 cm tall, 626-gram, yellow-green beryl home from Serenje, Zambia in the summer of 1999.*

Jeff Scovil photo

ultramafites that were subsequently metamorphosed to produce the schist and gemstones.

Aquamarine is exploited in several localities in Zambia: in the **Lundazi** area (adjacent to Malawi); at **Kapili Kesa**; **Kamanjiva;** and **Kaingu**. Aquamarine was mined in this region in the 1940s and 1950s. Later exploration revealed that the beryl/aquamarine pegmatites extended beyond Lundazi into the **Ndola**, **Namwala**, **Mpika** and **Serenje** districts.

Zimbabwe

Beryl is common in northern Zimbabwe and has been mined extensively through efforts ranging from large-scale operations to one-man diggings. About 160 pegmatites have been worked, and these have each produced more than 10 tons of beryl.

Beryl occurs in the **Filabusi**, **Masvingo**, **Save** and **Mberengwa** regions. While beryl may be the prime mineral mined, it is generally extracted as a byproduct while mining other commodities such as cassiterite, feldspar and tantalite. Some notable beryl pegmatites are **Cagal** in the Kariba district, which produced 479 tons of beryl between 1960 and 1974; **Dungusha** in the Hurungwe district, where gem-grade quartz, aquamarine, topaz and mica were also produced; and **St. Ann's** in the Hurungwe district, worked from 1953 to 1971. St. Ann's produced 1,268 kilograms of aquamarine, 1,006 kilograms of gem tourmaline, 936 kilograms of topaz, 14 kilograms of amethyst and 304 kilograms of mica. The **Good Days mine** in the Mutoko district yielded 315 kilograms of beryl between 1953 and 1972. During that same period, the mine also produced 2.07 tons of tantalite, 0.93 tons of cassiterite, 0.07 tons of columbite, 1,133 tons of spodumene, 580 tons of lepidolite and 91 tons of feldspar (Bartholomew, 1992).

Between 1962 and 1966, the **Flame Lily mine** in the Insiza district produced 12.46 tons of beryl, molybdenite, emerald, talc and powellite. At **Bepe Hills**, 90 km east of Mutare, beryl crystals 1.5 meters long and 60 centimeters wide were found. In 1956 a solid mass of beryl measuring 6x6x3 meters was uncovered at the **Rabbit's Warren mine** north of Mutoko. An unusual black variety of beryl has been described from the **Filabusi** area.

Aquamarine is found in pegmatites in the **Mwami** region. Stones with good quality color come from this area and to a lesser degree, from the northeastern regions of the country. Some of the finest yellow, gem-quality beryl comes from the **Green Walking Stick** deposit. Heliodor has also been found at the **Baboon Hill** and **Bul Bul claims** in the Mwami district. Gem quality aquamarine and heliodor is found in the **Hurungwe**, **Mutoko** and **Pfungwe** areas. Pink beryl is found at the **Pope claims**, Goromonzi district.

Zimbabwe is famous for emerald, which is found at **Mweza** (Sandawana), at **Odzi**, east of **Masvingo**, and also east of **Mutare**, north of **Gutu**, southeast of **Makuti** and finally southeast of **Filabusi**.

The most famous emerald deposit in Zimbabwe is **Sandawana** in the Mberengwa district. This world-famous deposit was discovered in 1956. Emerald is found in the southern Mweza greenstone belt in biotite and tremolite schist that encases small pegmatite veins. Sandawana emerald often has inclusions of other minerals including actinolite and cummingtonite, both members of the pyroxene group, as well as albite and fluorapatite. Apart from the famous Sandawana mine, there are 15 other emerald deposits in a 40-kilometer zone. These deposits are also associated with the Mweza greenstone belt rocks. There are several other emerald deposits in Zimbabwe although none are of the same scope as Sandawana. The **Machingwe** deposit located northeast of Sandawana has yielded good stones. At the **Chikwanda** prospect (Gutu district), over 17 kilograms of emerald was mined from 1962 to 1977; the largest specimen from the deposit weighed 160 carats. At **Brentwood** in the Masvingo district, emerald and beryl were mined in addition to chrysoberyl.

Zimbabwe

Above: *Golden beryl, 5.2 cm, from the Green Walking Stick mine; Desmond Sacco collection*

Below: *A typical hexagonal beryl crystal from St. Ann's mine, this Museum Afrika specimen is 7.2 cm tall.*

Both photos by the author

Brazilian Beryl

Luiz A. Menezes of Belo Horizante, Minas Gerais, Brazil is a long-time mineral dealer specializing in fine Brazilian minerals, particularly pegmatite species.

Brazil has long been a major source of spectacular beryl specimens. In fact, this South American nation was the 20th century's top producer of aquamarine and morganite and was the world's second largest producer of emerald. The color, shape and luster of Brazilian aquamarine crystals rank them among the finest and most sought-after mineral specimens in the world.

In Brazil, the different color varieties of beryl tend to occur in different pegmatite districts, with aquamarine and heliodor occurring in pegmatites that are in or close to the intruding granite; goshenite and morganite are found, together with colored tourmaline, in pegmatites that are situated several hundred meters to a few kilometers from the granite intrusion in the surrounding metamorphic rocks (see graphic below)—a schist in most Brazilian localities. Emerald, while sometimes deriving beryllium from the intruding granites, does not form in pegmatites, which are igneous, but instead tends to form in metamorphic settings.

Pegmatites that produce aquamarine can also host clear, light yellow, light blue or greenish topaz, schorl, magnetite and thin hematite blades that tend to form inside of thick mica plates, but these pegmatites never carry translucent tourmaline. A thick, rose quartz core may be present in the pegmatite, which may also contain smoky, tri-pyramidal quartz (one termination narrow, the other wide with prism faces that are obliquely inclined) that seldom achieve lengths of more than few decimeters.

In Brazil, long prismatic beryl crystals—aquamarine and heliodor—are generally found in zoned pegmatites that exhibit different grades of evolution from simple to intermediate to complex. Simple pegmatites tend to be narrow (usually less than 3 meters wide), and albite replacement bodies are absent. Intermediate pegmatites are up to 5 meters wide with few albite-filled bodies in which microcline was replaced by albite during the later stages of the pegmatite's formation. Complex pegmatites are sometimes more than 10 meters wide, albite replacement bodies are common, and large phosphate masses (triphylite and secondary alteration minerals) can be present, along with some lithium silicates (spodumene and lepidolite).

Tabular, colorless pink to orange crystals

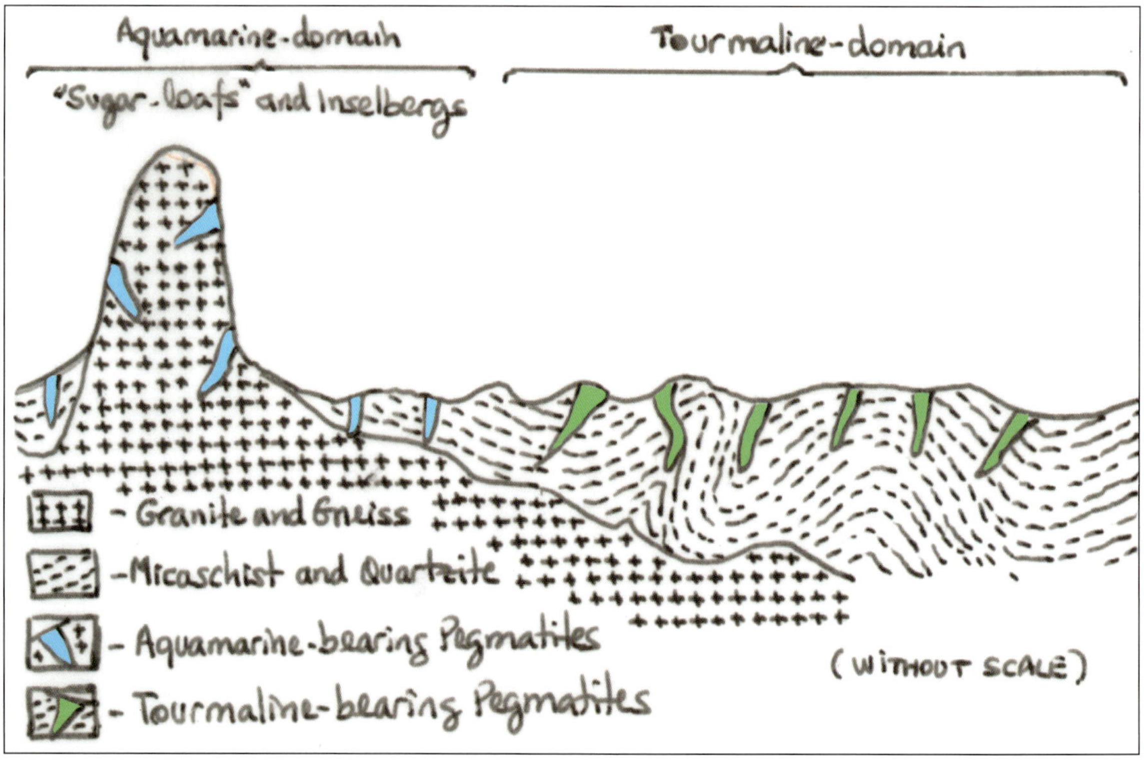

General position of Brazil's aquamarine and tourmaline-bearing pegmatites in relation to the granite source; graphic by the author after Cassadanne, (1996)

Beryl
12.6 cm high
Teófilo Otoni
District, Minas
Gerais, Brazil
Collection Bill
Larson; photo
Jeff Scovil

This fabulous 11.8 cm wide, gemmy etched beryl group is from the Medina mine in Minas Gerais.

Collection Scott Wallace; photo Jeff Scovil

—morganite and goshenite—tend to be found in even more complex pegmatites, with an abundance of albite replacement bodies, greater quantities of lithium silicates and pockets containing colored tourmaline. Within morganite-goshenite-tourmaline-bearing pegmatites, rich mineralized zones, with pockets relatively close to one another, are sometimes found. Such zones can sustain a steady production of gems for a long period of time. The **Cruzeiro mine**, for example, produced from the 1970s through the mid-1990s, the **Barra do Salinas** and **Santa Rosa mines** produced from the 1970s through the 1980s, and the **Pederneiras** from the 1980s into the mid-1990s and again beginning in 2000.

Pegmatites that produce aquamarine and heliodor in rich pockets on the other hand, tend to be more isolated from one another and zones of abundant pockets do not exist. Even after large pockets are found, such as at **Jaquetô** in 1990 and **Medina** in 1998, years of subsequent mining resulted in no further significant finds.

Aquamarine and Heliodor

Brazil's major sources for aquamarine and heliodor are the Oriental Pegmatite Province, the Northeastern Pegmatite Province and the pegmatites of northern Goiás. Goshenite and morganite are generally found in northeastern Minas Gerais; emerald comes from the schists of Bahia, Minas Gerais and Goiás.

Oriental Pegmatite Province

The vast majority of Brazil's finest mineral

Map of the

Index of Aquamarine Districts/Mines

1. Teófilo Otoni-Catugi-Padre Paraíso District
- *1.1 Catugi-Padre Paraíso*
- *1.2 Coroa de Ouro*
- *1.3 Chuchu*
- *1.4 Ariranha*
- *1.5 Gavião*
- *1.6 Avião*

2. Medina-Pedra Azul-Almenara District
- *2.1 Serrinha*
- *2.2 Blocos*
- *2.3 Manoel Silva*
- *2.4 Laranjeiras*
- *2.5 Ilha Alegre*

3. Jaquetô-Juerana District
- *3.1 Jaquetô*
- *3.2 Juerana*

4. Mimoso do Sul District
- *4.1 Concordia*
- *4.2 São Domingos*
- *4.3 Usina*

5. Center West Espírito Santo District
- *5.1 Pancas*
- *5.2 Pau Gigante*
- *5.3 Várzea Alegre*
- *5.4 Malacosta*
- *5.5 Rio Ponte*
- *5.6 Forno Grande*

6. Santa Maria de Itabira-Ferros District
- *6.1 Ponte da Raiz*
- *6.2 Posse*
- *6.3 Jatobá*

7. Conselheiro Pena-Galiléia-Resplendor District
- *7.1 Itatiaia*
- *7.2 Sapucaia*
- *7.3 Boca Rica*
- *7.4 Sapucaia do Norte (heliodor)*
- *7.5 Urucum (morganite)*
- *7.6 Jaîro (goshenite)*
- *7.7 Vista do Rio (morganite)*

8. South Bahia District
- *8.1 Morro da Glória*

9. Isolated Mines
- *9.1 Morro do Frade*
- *9.2 Bananal (morganite)*
- *9.3 Paíneiras (morganite)*
- *9.4 Cercadinho*
- *9.5 Veadinho (goshenite)*
- *9.6 Alvarenguinha*
- *9.7 Generosas*

Emerald Mines

E1. Belmont
E2. Capoeirana
E3. Brumado
E4. Anagé

Oriental Pegmatite Province

Index of Tourmaline Districts

10. Coronel Murta-Virgem da Lapa
Morro Redondo-Barra do Salinas-Xanda-Limoeiro-Manoel Mutuca mines

11. Itinga Araçuaí
Urubú-Maxixe-Pirineus-Poço D'Anta-Baixão-Laranjeira mines

12. Santa Rosa
Santa Rosa-Mutuca mines

13. São José da Safira-Marilac
Cruzeiro-Pederneiras-Chiá-Aricanga-Veadinho mines

14. Governador Valadares
Golconda-Escondido-Olho de Gato mines

15. Consolheiro Pena-Divinodas-Laranjeiras
Jonas-Sapo-Formiga-Pedra Alta-Pomaró mines

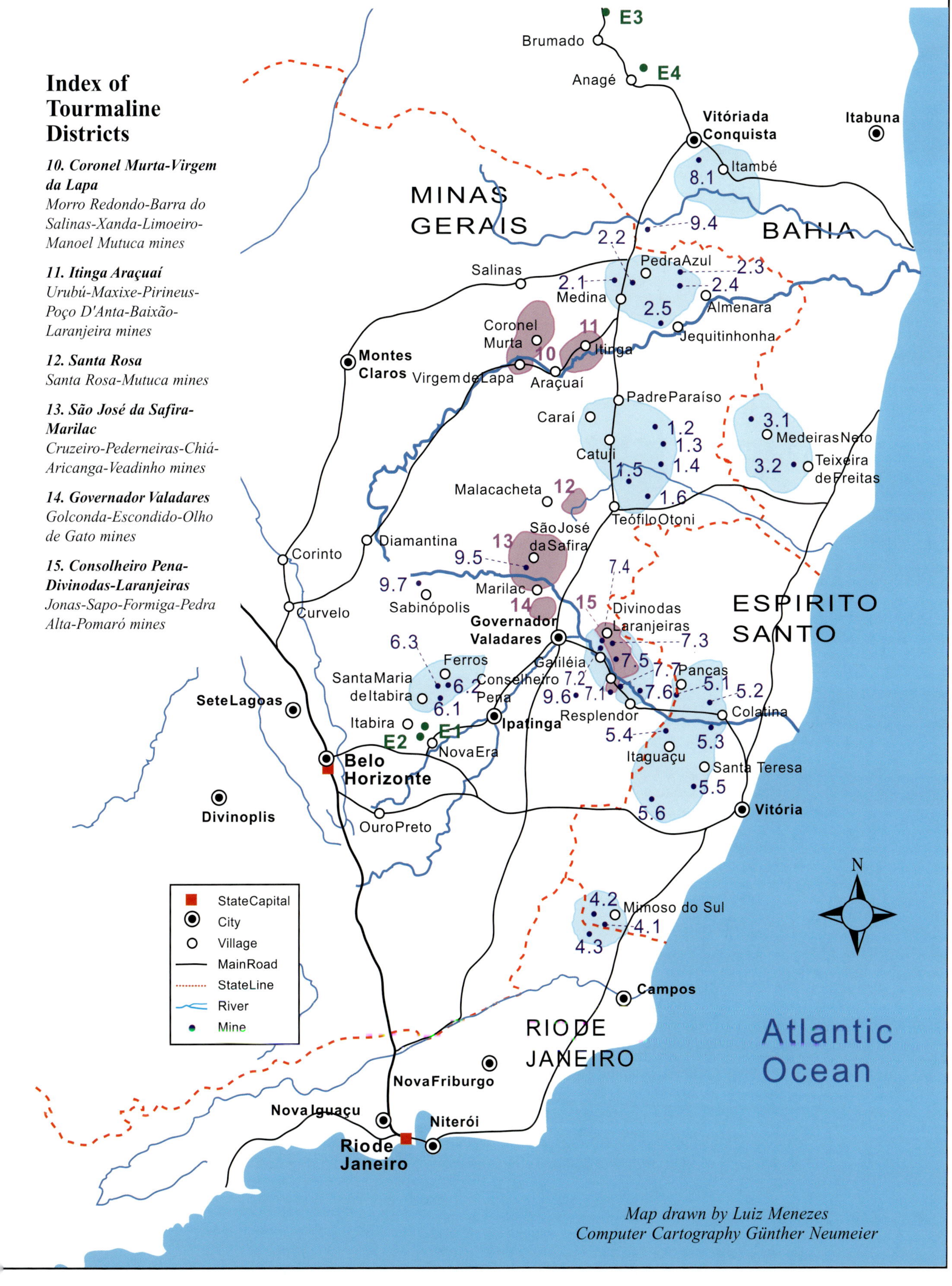

Map drawn by Luiz Menezes
Computer Cartography Günther Neumeier

Right: Back in 1998, an outcropping of the Serrinha pegmatite deposit near Medina in Minas Gerais produced some outstanding aquamarine specimens including this 8.8 cm tall crystal now in the collection of Andy Seibel.

Middle Right: *Italian dealer Riccardo Prato brought a wonderful lot of golden beryl crystals, including this 7.3 cm long specimen, back from the Abil Preto mine near Padre Paraíso in Minas Gerais.*

Both photos Jeff Scovil

specimens and gem rough come from the Oriental Pegmatite Province, which encompasses northeastern and central Minas Gerais, southern and southeastern Bahia as well as southern and western Espírito Santo. The province's most important districts, including the significant tourmaline deposits, are indicated on the map on page 87. The pegmatites of the Oriental Province formed between 650 and 450 millions of years ago and are related to extensive intrusive granitic bodies. There are nine major districts within this province.

Teófilo Otoni-Topázio-Catugi-Padre Paraíso-Caraí

The Teófilo Otoni district covers an area 100 kilometers long by 50 kilometers wide. This district is characterized by granitic inselbergs. The area has suffered intense weathering, and most of the area production has come from secondary colluvial, eluvial or alluvial deposits.

Beginning about 5 kilometers south of Catugi and extending north to Padre Paraíso, an 800 square kilometer area within the Teófilo Otoni district has proven highly productive for beryl. North of Padre Paraíso, secondary chrysoberyl deposits increase while the occurrences of beryl decline sharply. The more famous mines in this region are concentrated around the village of Ponto do Marambaia. These mines include **Lavra do Rex**, **Mucaia**, **Murundu** and **Córrego do Felipe**, all of which exploit secondary deposits of aquamarine. In fact, when the primary pegmatites were located beneath the secondary deposits, attempts to find aquamarine *in situ* have generally been unsuccessful.

Fine heliodor crystals have been found in this district. One 1999 find from a mine near Padre Paraíso uncovered gemmy, surface-etched, deep golden, prismatic beryl crystals with pale blue terminations that reached more than 16 centimeters in length. The district has also produced giant, colorless topaz crystals up to 50 centimeters long. Some of the deposits in the Teófilo Otoni such as the **Pica-Pau mine** also produce phenakite crystals (Cassadanne, 1985).

The most famous aquamarine found in this district was a deep blue crystal, weighing 34.7 kilograms; 33 percent of the crystal was of cutting grade. The crystal came from **Gavião**, an eluvial deposit near the city of Topázio. The crystal was found in 1954 and was named "Marta Rocha" for a marvelous woman from Bahia who had been named "Miss Brazil" a few months before. Soon after the beryl was found, Marta

Rocha became the first runner-up in the Miss Universe pageant. It was said at that time that only thing as nice as the blue of that aquamarine crystal were the blue eyes of Marta Rocha! After this spectacular find, hundreds of people gathered in the area and moved a good deal of soil, but nothing else was found.

Medina - Pedra Azul - Jequitinhonha - Almenara

Granitic inselbergs are also an important topographical feature of the 100-by-50 kilometer Medina pegmatite district. Well-known deposits in this district include **Serrinha** and **Blocos**, both near Medina; as well as **Manoel Silva**,

Laranjeiras and **Fortaleza**, all situated between Pedra Azul and Almenara.

The **Serrinha deposit** is located afew kilometers north of the city of Medina. Some 15 outcroppings in this area have produced important quantities of gem-grade aquamarine. In 1998, a small unnamed pegmatite produced some of the world's best aquamarine crystals: clear, blue prismatic crystals measuring up to 30 centimeters that were long and thin, with pencil-like terminations.

In 1998, the **Manoel Silva mine** in Almenara produced several hundreds of kilos of nice deep blue, highly etched aquamarine crystals. More than 100 kilograms of the material recovered was high quality, partially to completely clear pieces that were naturally etched into some wild shapes. One of the most dramatic pieces in this lot, a 22.5-centimeter high specimen dubbed "Ice Dragon," is currently in the collection of the Houston Museum of Natural Science.

Located near the city of Jequitinhonha on an island in the Jequitinhonha river, the well-known **Ilha Alegre mine** has produced some wonderful specimens, including a 65 kilogram gem aquamarine crystal found in 1972. The specimen was named "Cachacinha" for the miner, who apparently had an affinity for "cachaça," the popular Brazilian schnapps that is the product of the fermentation and distillation of sugar cane.

***Above:** Roz and Gene Meieran own this 11.3 cm tall crystal from the Medina mine in Minas Gerais.*

Jaquetô - Juerana

Several mostly eluvial and colluvial deposits in the southeastern portion of Bahia have produced large quantities of gem aquamarine. In 1990, a pocket in **Jaquetô** was found in a small pegmatite. This major pocket contained approximately 100 kilograms of very fine, clear, blue-green single prismatic crystals. The prism faces exhibited rectangular negative corrosion figures and pencil-like terminations.

Mimoso do Sul

The Mimoso do Sul pegmatite district is generally situated south of Espírito Santo but includes one small mine on the northern border of the state of Rio de Janeiro. **Concordia**, the

***Left:** A pocket in Jaquetô in the Brazilian state of Bahia produced this 10.6 cm long doubly terminated gem beauty, which is in the collection of Cliff Vermont.*

Both photos Jeff Scovil

district's major mine, exploits the primary pegmatite; in 1990 a very interesting pocket was uncovered there. It contained clusters of long, thin, pale blue aquamarine crystals with flat prism faces and pencil-like terminations. This deposit has also produced fine topaz crystals that range from colorless to yellow to green or blue and are often associated with smoky quartz. Colorless to pale pink phenakite crystals have also been recovered from the pegmatite.

West of Espírito Santo

Virtually all of the production from this district, which includes the cities of Fundão, Santa Tereza, Itaguassú, and Pancas, came from secondary, mostly alluvial, deposits. Within this district, there are also a number of chrysoberyl mines that have produced almost no aquamarine.

Conselheiro Pena - Galiléia - São Geraldo do Baixio - Resplendor

This area, 60 to 100 kilometers southeast of Governador Valadares in Minas Gerais, comprises several pegmatites that have produced large beryl crystals with clear "eyes" inside of them. In 1992, a pegmatite in **Galiléia** produced an aquamarine crystal with a perfect basal termination that measured 1 by 0.5 meters and weighed 400 kilograms! I purchased the piece at the mine and it is now housed in the Kanagawa Prefecture Museum near Tokyo. Several beryl-bearing pegmatites in this district, including **Sapucaia**, **Cigana**, **Boa Vista** and **Boca Rica**, also produce a variety of phosphates such as triphylite, huréaulite, vivianite, amblygonite, ludlamite, zanazziite and frondelite.

The most important aquamarine producer in this district is the **Itatiaia pegmatite**, located 10 kilometers from the famous **Jonas mine**, in Conselheiro Pena. Itatiaia has consistently produced large quantities of etched beryl crystals, ranging from blue to bluish green as well as small quantities of etched heliodor. In December 2003, miners opened a large pocket containing tabular green beryl crystals up to 20 centimeters long, perched atop gray muscovite and albite.

***Above Left**: 8 cm high; Concordia mine, Mimoso do Sul, Minas Gerais; collection Cal & Kerith Graeber*

***Above Right**: 10.5 cm high; Minas Gerais; collection Bob Whitmore*

Both photos Jeff Scovil

Santa Maria de Itabira - Ferros - Rio Piracicaba

The district of Santa Maria de Itabira is situated in central Minas Gerais, about 100 kilometers northeast of Belo Horizonte. The best known mine in this area is **Ponte da Raiz**, in Santa Maria de Itabira. Beryl from Ponte da Raiz was initially pulled from the eluvium and later from the primary pegmatite. The mine has produced coarse crystals with an intense blue color and good gem yield. Other important deposits in the district are **Posse** and **Jatobá**. The district also produced very large (up to 100 kilograms) clear topaz crystals.

Southern Bahia

In southern Bahia near the cities of Itambé, Vitória da Conquista and Encruzilhada, several pegmatites have produced industrial feldspar and columbite as well as large beryl crystals with clean "eyes" at their cores. The most famous mine in the area is **Morro da Gloria**.

Other Districts in the Oriental Province

The Oriental Pegmatite Province is home to numerous other pegmatite regions. In Minas Gerias alone, there are many well known regions such as **Coronel Murta** (the Morro do Frade mine produced intense blue gem rough), **Virgem da Lapa**, **Itinga**, **Araçuaí**, **São José da Safira**, **Santa Maria do Suaçuí**, **Peçanha**, **Governador** *Valadares*, **Chonin**, **Alvarenga** (Alvarenguinha mine), **Belo Oriente** and **Sabinópolis** (Generosas mine). While all are home to important, productive deposits, most have only produced beryl crystals with gem-grade cores.

Northeastern Pegmatite Province

The Chapada da Borborema mountain region, which encloses the south-central portion of the state of Rio Grande do Norte as well as the north-central portion of Paraíba, is a complex of metamorphic rocks that is estimated to be about 500 to 600 million years old. These rocks, which extend through northeastern Brazil from Rio Grande do Norte to Paraíba and into Ceará, host many zoned pegmatites. These pegmatites are not highly evolved and may contain centimeter to meter-sized beryl crystals that are embedded in quartz or microcline. Since World War II, the area has been mined extensively for industrial beryl and tantalite. The mines have on occasion produced beryl with gemmy sections as well as decorative specimens of blue-green beryl embedded in matrix.

Since the late 1980s when the market for industrial beryl was depressed by the discovery of rich bertrandite deposits in North America, most of the mines in this province have been closed. One mine that must be mentioned is located southwest of Rio Grande do Norte in Tenente Ananias. This mine has produced notable quantities of medium to good color gem rough.

Goiás

In the northern section of the state of Goiás,

This well known piece was in the collection of Herman Bank and is pictured in his books, Precious Stones and Minerals *(pg. 55) and* From The World of Gemstones *(pg. 100). It was sold to Joseph Freilich and was pictured on the back cover of the* Mineralogical Record *(31:1).*

The specimen is reportedly from the Marambaia district, Padre Paraíso in Minas Gerais and measures 14.3 cm in height.

Collection and photo Stuart and Donna Wilensky

The Urucum mine produced this deep pink, tabular beryl crystal on matrix. The overall specimen is 40 cm wide. Collection and photo Adalberto Giazotto

the Serra da Mesa pegmatite district has produced medium blue as well as pale pink beryl crystals, though there have been no reports of important finds in recent years.

Morganite and Goshenite

Northeastern Minas Gerais

Most of the pegmatites that host morganite and goshenite are located in the northeastern portion of Minas Gerais. These complex pegmatites are zoned, and the beryl is generally found inside cavities in the substitution bodies. Crystals are often perched on albite or microcline, muscovite and/or lepidolite, schorl and/or gem colored tourmaline, making for some fabulous specimens.

In 1973, the **Urucum mine** in Galiléia yielded a pocket containing some of the best morganite ever found: intensely pink, tabular clear crystals that measured up to 30 centimeters wide. These magnificent specimens formed atop albite and were pierced by dark green tourmaline crystals. In 1968, a huge pocket of kunzite had been found at the same mine, and in 1988 another giant pocket was opened that contained more than 100 kilograms of high-grade kunzite, 100 tons of giant quartz crystals, a few nice purple apatite crystals and tiny, clear stokesite. Unfortunately, however, the pocket did not contain morganite.

Since 1983, the **Vista do Rio mine**, located on the left bank of the Doce River in Conselheiro Pena, has been producing fine tabular pink morganite crystals on albite in association with lepidolite and schorl. In 1989, production of high-grade morganite from the mine was reportedly some 300 kilograms. The best pocket was opened in 1997 and contained one superb crystal on albite that measured 20 centimeters in diameter.

The **Paineiras mine** in Coronel Murta is located 2 kilometers from the Morro Redondo mine and 2 kilometers from the Barra do Salinas mine. All three mines work the same geological

***Above:** Beryl on Muscovite; 4.2 cm high;*
Jairo mine, Resplendor, Minas Gerais
Collection Bill Mickols

***Right:** Beryl on Albite; 9.6 cm high;*
Jairo mine, Resplendor, Minas Gerais
Collection Keith Proctor

***Below:** Beryl and Quartz*
47 cm wide; Pederneira mine, Minas Gerais
Collection Rice Museum

All photos Jeff Scovil

Above: *Goshenite crystal and 12.39 carat facetted stone*
Brazil; collection Michael Scott
Photo Harold and Erica van Pelt

Below: *Beryl*
25 cm in diameter by 30 cm high; Golconda mine, Minas Gerais
Collection and photo Adalberto Giazotto

formation. In the last few years, Paineiras has produced several pockets containing deep pink beryl crystals weighing up to 20 kilograms. These were partially to deeply etched and constituted valuable gem rough. A smaller quantity of nearly colorless material was irradiated to induce bright "sugar-cane green" color. A few excellent matrix pieces were also recovered.

Also in Coronel Murta, the **Bananal mine** has, since 1960, yielded several large pockets of pink and green beryl. In 1986, 500 kilograms of beryl crystals with salmon-colored centers and light green "rinds" was recovered from the mine. The largest crystal was 18.5 centimeters in diameter.

In 1987 and 1988, the **Jairo mine** in Resplendor produced some very nice, thin, tabular beryl crystals that were clean, bright, slightly greenish and stacked together in parallel growth. They were associated with long, thin, dark green tourmaline crystals; the largest specimen weighed 2 kilograms and measured 15 centimeters. In 1988, the same pegmatite produced a huge pocket containing several hundred kilos of high-grade gemmy, pink spodumene.

In the early 1980s, the **Veadinho mine** in Marilac produced very fine tabular, colorless beryl crystals in association with green tourmaline. The best piece measured 7 centimeters in diameter and was sitting on a microcline crystal. The specimen is now in the Houston Museum of Natural Science collection.

In addition to those mentioned above, many other pegmatites have produced pink, colorless and green beryl. These include **Barra do Salinas** in Coronel Murta, **Maxixe** and **Sebastião Dutra** in Itinga, **Pederneiras** and **Chiá in São José da Safira**, as well as **Golconda** and **Jacó** in Governador Valadares.

Emerald

Brazil is second only to Colombia as the world's largest producer of emerald. Unfortunately with the exception of Brumado, all of the mines produce emerald crystals that are embedded in the surrounding matrix. The crystals are therefore not lustrous and few good collector specimens are culled from the production.

Bahia

Emerald was first found in the **Pindobassú area** in the late 1960s. During the peak of

Beryl
6.6 cm high;
Visto do Rio mine, Conselheiro Pena, Minas Gerais, Brazil
Collection Sandor Fuss; photo Jeff Scovil

production in the 1970s to mid-1980s, hundreds of people were working numerous narrow underground operations. These mines typically consisted of a narrow shaft up to 100 meters deep. Workers were lowered into the mines in drums. They worked the horizontal adits that followed the narrow emerald-bearing veins. These operations were later organized into the **Carnaíba** and **Socotó mines**.

In addition to emerald that sometimes forms an interesting association with molybdenite, Carnaíba produced flattened alexandrite crystals that exhibited "V-twin" or sixling habits. Alexandrite crystals measure up to 5 centimeters, and though they are fairly dark, they do show a marked color change. The Socotó mine produced large, rough phenakite crystals.

This 7.7 cm high emerald group from Brumado in the state of Bahia, Brazil, was formerly in the collection of Gilles Emringer and is currently owned by Roz and Gene Meieran. Photo Jeff Scovil

Brumado, known for producing spectacular uvite in addition to magnesite, dolomite, hematite, Japan-law twinned quartz and sellaite, has also produced Brazil's best emerald specimens. The "mine" is indeed a group of several extensive open pits that are exploited to produce some millions of tons per year of magnesite ore. Magnesite is used on site to make refractory bricks. Brumado is the largest producer of factory bricks in the southern hemisphere. Unfortunately, the company has fiercely prohibited specimen collecting by mine workers, and many fantastic specimens are thus destroyed. In 1997, an outstanding lot of emerald was found at the **Pombas pit**. The find included 5 clusters of deep green, sharp bright crystals, measuring up to 10 centimeters. The clusters consisted of one large complete crystal with numerous tiny crystals growing perpendicular to the prism face of the central crystal. Emerald has also been found at Brumado's **Pirajá pit**.

In the state of Bahia, **Salininha**, **Anagé** and **Fazenda dos Pombos** have also produced some emerald.

Goiás and Minas Gerais

The **Santa Terezinha emerald deposit** in Goiás was discovered in 1981, and it soon attracted people from Carnaíba, where the veins were becoming too deep to be economically worked. In Santa Terezinha, emerald is still extracted through deep, narrow underground operations. The resulting crystals are embedded in schist, quartz or calcite.

Discovered in 1978, the **Belmont deposit** in Itabira, Minas Gerais was initially worked manually. Since 1981, however, the operation has become more mechanized and involves open-pit mining of the emerald-bearing schists, followed by washing, screening and manually separating the emerald from the gravel.

Also in the state of Minas Gerais, the **Capoeirana mine** in Nova Era was discovered in 1988. Soon after its discovery, the deposit was the focus of an intense and sometimes bloody dispute over the ownership of the land and mining rights. Finally in 1992, a miners' cooperative was founded, and it took over the operation. Similar to some of the above-described mines, Capoeirana's narrow mineralized veins are accessed by adits and shafts. Capoeirana has produced good gem emerald crystals weighing up to 800 grams.

Mara Rosa, **Fazenda das Lages**, **Porangatu** and **Pirenópolis** in Goiás and **Santana dos Ferros** in Minas Gerais have also reportedly produced emerald, as have **Pela Ema** in Tocantins and **Tauá** in the state of Ceará.

Enormous Potential

Brazil's enormous 20th century production of beryl was due to a combination of several factors, including the existence of virgin, rich secondary deposits; the availability of cheap labor; and a lack of environmental constraints. As the new century begins, the situation has changed dramatically as many of the secondary deposits in the known districts are being exhausted. In addition, labor costs are on the rise and environmental and safety standards are becoming stricter and more expensive to implement. Nevertheless, there is enormous potential for the discovery of new pegmatites, and large portions of the primary deposits of most of the known localities remain untouched. It seems there will always be independent miners willing to take a chance and mine those reserves. Brazil will thus likely continue to be an important producer of beryl over the next several decades.

Bolivia and Colombia

Bolivia *by Alfredo Petrov*

Bolivia is the only South American country, outside of Brazil and Colombia, with a notable beryl production. That production, however, pales in comparison to the gemstone powerhouses. Large quantities of beryl are found in several pegmatite districts in the eastern lowlands of Bolivia, which is a geologic extension of the Pre-Cambrian Brazilian Shield. These districts have not been explored nearly as thoroughly as those on the Brazilian side of the border.

The pegmatite district east of the town of **San Ramon**, in Santa Cruz Department, was originally worked for large sheets of mica in the 1930s and 40s. After WWII, when beryllium became a desperately needed strategic material for the nuclear industry, American mineralogist Dr. Richard Gaines was sent to Bolivia to investigate reports of beryl from there. He spent a few uncomfortable weeks traipsing through the then roadless jungle on horseback and was quite surprised by how prolific beryl was in the abandoned mica mines of the San Ramon district. Beryl occurs here as opaque white to pale blue or greenish crystals up to a meter or more in size. **Mina La Bella**, **Mina La Verde** and other pegmatites in this district produced about 1,500 tons of beryl. In spite of the drop in demand for industrial beryl, the same mines are still being worked for tantalum ore. Gem prospectors pick over the remaining beryl looking for gemmy patches that occasionally yield faceted aquamarine up to 2 carats.

Other pegmatite districts in Santa Cruz Department also produce beryl, most importantly the **Los Patos district** near the city of San Ignacio de Velasco, where white to pale yellow crystals can reach 3 meters in size. These are often accompanied by muscovite and tantalite in a matrix of kaolinized feldspars. Dark green, non-gemmy emerald has been reported from a couple of areas in eastern Santa Cruz. Though these are of little value, hopeful prospectors are still quite secretive about the exact localities!

No beryl has ever been found in the vast mountain ranges of the Bolivian Andes. A reported find of "aquamarine" in the Andes of La Paz Department turned out to have been misidentified apatite.

King of Green
Colombia remains the world's leading emerald producing nation. This 2 cm high, doubly terminated crystal on calcite is from the Muzo mine.
Jeff Collins collection; Jeff Scovil photo

Colombia *by Rupert Hochleitner*

Colombia remains the world's number one producer of emerald, producing not only the most stones, but also the best and biggest. Nearly all of the large, high-quality cut emeralds on the market come from Colombia.

The Colombian localities can generally be divided in two regions. Localities in the western zone include **Coscuez**, **Yacopi**, **Muzo**, **La Pita** and **Penas Blancas**. In the eastern zone the major producers are **Chivor**, **Gachala** and **Macanal**.

The geologic situation is similar in both regions: calcite, dolomite and feldspar dike, cleft and fissure fillings in schistose breccias contain hydrothermally formed crystals that are up to 10 cm long. Most, however, are far smaller. These crystals have few inclusions and tend to be very high in quality. Accessories of Colombian emerald include calcite, pyrite, fluorite, apatite, muscovite, tourmaline, barite, and rarely parisite or euclase.

The Colombian emerald deposits have been covered in much greater detail in *extraLapis English No. 2: Emeralds of the World.*

Beryl Red and Green

William "Skip" Simmons and Alexander Falster on the most valuable of the beryl color varieties

Emerald is probably the most valuable beryl variety. For that reason, the debate over its definition has been raging for generations. It is generally accepted that a deep grass-green beryl is an emerald if it derives its color from the presence of chromium or vanadium. Interestingly in addition to chromium or vanadium, synthetic emerald requires a trace amount of nickel in the melt in order to meet many people's definition of "emerald green," but nickel is rarely present in natural emerald.

Emerald occurs notably in two major types of environment: hydrothermal veins in dark limestones such as those in Colombia near Muzo and Chivor; and in biotite schists (Habach Valley, Rauris, Austria).

Beryl variety Emerald with Calcite
5 cm; Coscuez, Boyaca, Colombia
Collection Bill Larson; photo Jeff Scovil

When emerald is found in proximity to a pegmatite, it is commonly close to the contacts with a chromium or vanadium-bearing country rock, as neither chromium nor vanadium are typically found in elevated concentrations in granitic pegmatites. These elements typically represent a contaminant of the surrounding country rock.

Morphologically similar to emerald, the red variety of beryl has been marketed under various names including the oxymoron "red emerald," "raspberyl" and the now obsolete "bixbite." Red beryl crystals have a remarkably intense raspberry-red color, probably due to trivalent manganese.

The morphology of red beryl and emerald is generally simple, in many cases only including the basal pinacoid {0001} and the hexagonal prism $\{10\bar{1}0\}$. Beryl in general most commonly takes these two forms, but emerald and red beryl exhibit them almost exclusively, though some crystals may exhibit small additional forms such as dipyramids.

Red beryl is a low-alkali variety of beryl, which tends to contain very little water. Like red beryl, the beryl-group species pezzottaite and the beryl variety morganite derive their color from the presence of trivalent manganese. Though related to beryl, pezzottaite is a different mineral species with a chemical formula distinct from that of beryl of any color. Red beryl is different from high-alkali morganite in that morganite tends to be tabular and forms in pegmatites. Red beryl is generally prismatic and is essentially restricted to topaz rhyolites, for example in the Wah Wah Mountains, Beaver County, Utah; the Thomas Range, Juab County, Utah (Shigley, 2003); the Black Range in New Mexico and reportedly near San Luis Potosi in Mexico (Sinkankas, 1981). It occurs in lithophysal cavities and fractures as well as within solid rhyolite.

The classic locality for red beryl is the Wah Wah Mountains, which was discovered in 1958 by Lamar Hodges. In January 1977, brothers Ed, Rex and Bob Harris bought the claim for just under US$ 50,000, thinking it would be fun to mine the property. Fun it has been, not to mention profitable; since 1977, the Ruby Violet claim has produced gem rough valued at more that 35 million US dollars.

The most intensely-colored varieties of beryl, emerald and red beryl are both highly desirable as gemstones and as mineral specimens.

Beryl variety Red Beryl
The large crystal is 1 cm wide.
Searle Canyon, Thomas Range, Utah
John Holfert collection; Jeff Scovil photo

Beryl variety Red Beryl
main crystal 5.5 cm high
Ruby Violet claim, Beaver County, Wah Wah Mountains, Utah, USA
Collector's Edge specimen; Jeff Scovil photo

Grading and Pricing Gem Beryl

Chemist, graduate gemologist and full-time lapidary John Bradshaw has been the Curator of Gemstones at the Harvard Mineralogical Museum for 20 years.

While the term "beryl" is not well known to the general public, a few of the many varieties of this species are among the oldest and best known gemstones in history. Perhaps, the best known of these varieties is emerald. Also well known is aquamarine, which is typically found as a greenish-blue stone. Beryl, as with all colored stones, is evaluated, graded and priced using the same characteristics as those used to evaluate diamond: color, clarity, cut and carat weight.

Color is based on a combination of hue, tone and saturation. According to the GIA's colored stone course, hue is a position of color on a color wheel. There are 31 hues used in the GIA grading system. Saturation is the intensity of that color ranging from brownish or grayish to vivid. Tone is the lightness or darkness of the color ranging from very light to very dark. A top-grade emerald, for example, would have a dominant green body hue with a blue secondary modifier; it would be medium in tone, with strong saturation.

This 4 cm long morganite gemstone was cut from Brazilian rough. Photo Harold & Erica van Pelt

Clarity is based on evaluating internal as well as the external characteristics. In most cases, the absence of inclusions will increase the value of a stone. GIA classifies transparent gemstones into types to recognize that all gems do not form with the same clarity in nature. Type I stones include those that are found in nature with relative regularity having no eye-visible inclusions (e.g. aquamarine). Type II encompasses gems that are typically found with some minor inclusions that may be visible to the eye (e.g. sapphire). Type III gemstones are generally found in nature with many eye-visible inclusions (e.g. emerald). Clarity of colored stones is described as eye-clean, slightly included, moderately included or severely included.

The cut of a gem can greatly influence its value. Well-cut stones can enhance the overall color grade and therefore its value. Conversely, poorly cut stones may greatly reduce a stone's value even if the color component is considered excellent. Factors considered in grading cut are brilliance, extinction, windowing, symmetry and finish.

I am often asked how much one should expect to pay for rough. Unfortunately, this question has no easy answer. Rough is generally sold in parcels that contain pieces of varying qualities. Some parcels might be quite close in quality while some vary greatly. Each gem species has its own set of optical properties and thus its own set of parameters as to what can or cannot be done. The cutter, therefore, evaluates color and clarity in a fashion similar to that of evaluating cut stones but in addition must consider orientation for cutting based on the shape of the rough to maximize size, proportion, color, clarity and beauty. Percentage yield may vary greatly from piece to piece within a parcel. Experienced cutters take all of these factors into consideration to determine whether the value of the rough is priced fairly or not.

Perhaps an even greater problem in today's marketplace is that there is a drought of rough of all types. In the past, when gem production was slow in one area, it was up in others, but that is not the current status of the market. Older deposits are not producing the quantities they once did. New deposits are few and far between. Since most rough is found in poor countries, many factors influence the supply of rough including the lack of sophisticated mining equipment, bad weather and political unrest. The recent decline of the US dollar against world currencies has also played a role, making it difficult for American dealers to buy rough. What little rough that is being found is rapidly bought by European or Asian firms at prices that are significantly higher than those the American dealer can pay if he expects to sell the finished gems at a profit.

Unfortunately there is no universal grading system for colored stones as there is for diamonds. A grader's experience and market knowledge determines how accurately he can grade and therefore price a stone.

For the purposes of the charts to the right, a grade of "A" will refer to an average quality, "AA" to above average quality and "AAA" to exceptional qualities. Commercial quality or "B" grade gems would be priced less than the "A" grade in each variety. In the charts that follow, values given estimate the current retail prices per carat in US dollars.

Emerald has been found in many countries worldwide but most commercially available material is from Colombia, Brazil and Zimbabwe. The color range can be very strongly bluish-green to very slightly bluish-green, green and slightly yellowish-green. Being a Type III gemstone, emerald typically has eye-visible inclusions. Common enhancements include oil and resin filling.

	A	AA	AAA
0.5 to < 2 ct	250-2,000	1,100-6,500	3,000-13,000
2 to < 4 ct	800-3,500	3,200-10,000	8,000-20,000
4 to < 5 ct	1,500-4,500	4,500-11,000	11,000-22,000

Aquamarine is found in commercial quantities in Brazil, Zambia, Nigeria and Madagascar. Colors can be blue, very slightly greenish-blue to very strongly greenish-blue. As a Type I gem, it is usually eye-clean. Heat is the most common enhancement that removes the green component yielding a blue gem.

	A	AA	AAA
1/2 to < 3 ct	30-125	100-500	250-1,200
3 to < 10 ct	60-250	150-600	500-1,200
10 to < 25 ct	125-250	250-650	650-800
25 ct +	125-225	225-400	400-700

Golden beryl or **heliodor** is most often found in Brazil, Ukraine and Madagascar. It ranges in color from an orange-yellow to yellow to greenish-yellow. Like aquamarine, it is also a Type I gem and is usually eye-clean. Heat and irradiation are the common enhancements for golden beryl.

	A	AA	AAA
1 to < 5 ct	50-80	60-100	80-125
5 to < 10 ct	75-100	100-150	150-200
10 to < 20 ct	100-125	125-200	200-500

Morganite is found predominantly in Brazil, although Madagascar has produced fine gems in recent years. Being a Type I stone, eye-clean stones are the norm. Heat is used to burn the brown or orange component from morganite yielding a pure pink color. Depending upon the strength of sunlight in a particular region, the same result can be obtained by leaving morganite in the sun for a week or so.

	A	AA	AAA
1 to < 5 ct	20-100	75-200	175-800
5 to < 20 ct	60-150	125-300	275-1000
25 ct +	75-150	150-300	300-600

Goshenite is found in Brazil, Russia and Maine. Like its other beryl cousins, it is usually eye-clean. No enhancements are known. It does not have a particularly large market due to its lack of color. Prices would generally range between $10 and $80 per carat in all sizes

Red Beryl is similar in many ways to emerald; it is almost always found with eye-visible inclusions. Oiling or resin filling is the enhancement of choice. Not enough material exists in the marketplace to establish firm pricing but can be estimated to range from a low of $1000 per carat to a high of $20,000 per carat. Exceptional stones of one carat or more would be considered near the higher end of the range.

Element Beryllium

Beryllium buff Lee Van Iderstine of Shelton, Connecticut and* Lapis *editor Michael Huber of Munich, Germany write a few words about the element beryllium and its industrial significance

French mineralogist Abbé René Just Haüy thought that beryl might contain an as yet unknown element. Based on that suspicion and the remarkable similarities that he noted between beryl and emerald, Haüy asked chemist Louis Nicolas Vauquelin (1763-1829) to look into the matter. Haüy's hunch proved correct: in 1798, Vauquelin announced that he had educed beryllium oxide from both common beryl and from emerald though he was unable to separate the element from its oxide.

Louis Nicolas Vauquelin (1763-1829) isolated the beryllium oxide from the mineral beryl and gave the new element the name "glucinium." Archive Ecole des Mines, Paris, France

Because of the sweet taste of the compound, Vauquelin named the element "glucinium" for the Greek word "glykys," meaning sweet. German chemist Martin Heinrich Klaproth (1743-1817) later duplicated Vauquelin's work, naming the element "beryllium." The French continued to use the term "glucinium" instead of beryllium until 1957.

Using the action of potassium on beryllium chloride both Friederich Wöhler (1800-1882) and Antoine A. Bussy independently isolated an impure form of the element beryllium in 1828. It was not until 1899, however, that Marie Alfred Lebeau (1868-1959) first isolated a pure form of the metal.

Beryllium is the second lightest metal after lithium. Pure beryllium does not occur naturally and is extracted primarily from bertrandite and beryl. The pure element is generally obtained through the electrolysis of a fused mixture of beryllium chloride and sodium chloride.

Pure beryllium is lustrous, brittle, silvery-white and rare: the Earth's crust is only 0.0006 percent beryllium. An alkali earth metal, beryllium has the atomic number of 4, an atomic weight of 9.012182 and a hardness of 6.5. Its density is 1.8 kilograms per liter, making it less than twice as dense as water and only two-thirds as dense as aluminum. Beryllium has the very high melting point of 1,287 degrees Celsius and a boiling point of 2,471 degrees Celsius. Even at red heat, beryllium is unaffected by air or water. Beryllium is a nonmagnetic, nonsparking metal; it conducts electricity and has a high capacity to absorb heat. The specific stiffness of metallic beryllium is much greater than that of other engineered structural materials such as aluminum, titanium and steel. Its properties make beryllium, though expensive, conducive to a number of industrial uses.

Element of the Cold War

A little more than 30 years after it was isolated as a metal, the element beryllium played an important role in the advancement of nuclear science when in 1932 British physicist James Chadwick (1891-1974) fired alpha rays at a sample of beryllium and found that the sample emitted a previously unknown subatomic particle that had mass but no charge—the neutron. For his discovery, Chadwick won the 1935 Nobel Prize for Physics. During World War II, Chadwick was made head of the British mission attached to the Manhattan Project for development of the atomic bomb.

Beryllium is unusual in that it does not absorb neutrons. It is thus used in nuclear weapons and reactors. In a nuclear warhead, energy is released from uranium when it is bombarded with neutrons. As the beryllium does not absorb but even repels neutrons, beryllium parts help to

A Word of Caution

Beryllium compounds are toxic to people. In spite of the experiments of early chemists, do not taste or ingest any form or amount of beryllium, as it mimics and replaces essential magnesium in the body causing illness or death. Inhaling beryllium dust or vapor can cause berylliosis, a condition that can severely damage the lungs, settle into the skeleton and can be fatal. Beryllium is not for human consumption.

maintain a higher neutron flux within the bomb. Not surprisingly, in the 1940s, the U.S. Atomic Energy Commission became the first major consumer of beryllium metals and beryllium oxides.

During the Cold War, beryllium held high strategic value, which led to the discovery and active mining of a number of beryl and bertrandite deposits. In fact, during the Cold War, emerald and alexandrite were merely byproducts of the Soviet's beryllium ore recovery efforts at the famous **Izumrudnye Kopi** near the village of Malysheva in the central Urals of Russia (see page 49).

Industrial Element

In the mid-1950s, beryllium was made available to industry and a variety of commercial applications were developed that employ pure metal and its alloys. Beryllium and beryllium alloys provide the raw material for many useful items in the aircraft, space exploration, electronics and telecommunications industries. Transparent under x-ray, beryllium is used in the windows of x-ray transmission tubes.

In 1926, M.G. Corson found that beryllium can be used to age-harden copper and nickel-copper alloys. The strength of copper is increased six-fold by the addition of two parts beryllium. Bringing more than just strength to the mix, when copper is alloyed with beryllium, the electrical and thermal conductivity of copper is improved; and copper becomes harder with better corrosion and fatigue resistance. Beryllium alloys are also remarkably elastic. Copper-beryllium alloys, which generally contain between .25 and 5 percent beryllium, are thus chosen for a variety of applications including computer main boards and processors, automotive electronics, plastic mold tooling, miniature electronic connectors and aircraft bushings.

Higher strength alloys tend to have moderate conductivity and offer the best wear resistance, fatigue strength and hardness available in the copper alloy system. More highly conductive alloys offer moderate strength and are ideally suited for heat dissipating modules, resistance welding components and other electrical applications.

Copper-beryllium alloys can be machined into intricate shapes. In fact, the machinability of age-hardened beryllium-coppers is comparable to that of many other copper-based alloys and is more machinable than stainless steels. Copper-beryllium alloys also offer dimensional stability that is far superior to that of any other copper

***Upper:** The one-time logo of the Brush-Wellman company cast here in pure beryllium; 9 cm high*
Collection Lee Van Iderstine; photo Jeff Scovil

*NGK Metals manufactures a line of beryllium-copper alloy "safety" tools such as the coal chisel pictured **middle**.*
Photo courtesy of NGK Metals

*Some high-quality clocks are equipped with beryllium alloy springs such as the one pictured **below**.*
Photo Kamber, Bucherer AG

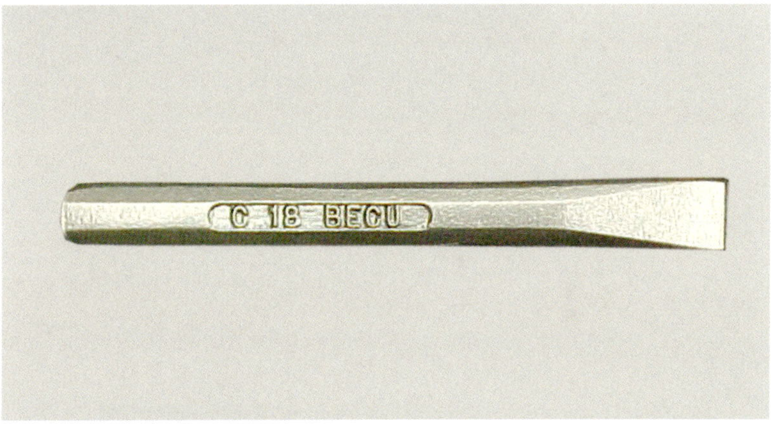

Beryl was at one time an important source for beryllium. Today, however, the most important ore is the mineral bertrandite.

Right: Open pits of the largest economic beryllium deposit in the world: Spor Mountain in Juab County, Utah.

based alloy. Beryllium-coppers are wear resistant in comparison to other copper-base alloys and to many ferrous alloys. The high strength of age-hardened copper-beryllium parts makes them capable of bearing loads far in excess of other copper based alloys and gives the added advantage of compatibility with high hardness ferrous alloy shafts.

Beryllium-coppers are nearly as corrosion resistant as nickel-silver alloys, but are not susceptible to hydrogen embrittlement. Tools made with beryllium copper are nonsparking, an important feature when working in volatile situations, such as oil and gas well drilling where the slightest spark could spell disaster. As beryllium is not magnetic, it is extremely useful in alloyed parts used for housing sensitive detectors affected by magnetic fields.

Beryllium-copper bearings are being tested in the engine rotors of electric cars. AC Propulsion has developed a prototype car that can go from 0 to 60 miles per hour in under 4 seconds and has a 300-mile range between charges. One of its secrets is the beryllium-copper bearing supporting the engine rotor. This bearing is highly wear-resistant, allowing higher rpms.

Nickel-beryllium alloys are used in high temperature applications such as in electrical appliances. Lockalloy is an aluminum alloy, which is used in airplanes, satellites and rockets. Beryllium oxide is an excellent heat conductor, with high strength and hardness. Beryllia ceramics are more thermally conductive than many other materials and are used in high power electronic circuitry, laser equipment and microelectronics packaging.

Northrop Grumman is currently working in conjunction with NASA on the James Webb Space Telescope, scheduled to launch in 2011. This new telescope is expected to enable scientists on Earth to see objects 10-11 billion light years away. The primary mirror of the James Webb Telescope will be made out of an optical-grade beryllium. The mirror, developed by Ball Aerospace Technologies Corporation, is beryllium-based because of the metal's lightweight, strength, stiffness and machinability. The mirror will be 6.5 meters in diameter and will be comprised of 18 smaller hexagonal mirrors. It will have to withstand temperatures less than 50 degrees Kelvin (-220 degrees Celcius.

Beryl is the primary beryllium ore extracted in Russia, but in the United States, bertrandite is the major ore for beryllium. Together, these two countries produce more than 95 percent of the world's beryllium supply. The United States alone produces about 65 percent. The primary source for beryllium ore in the United States is **Spor Mountain** in Juab County, Utah. Other beryllium-producing countries include Brazil, China, Argentina, Kazakhstan, Madagascar, Mozambique, Portugal and Zambia.

Literature

Alferova, M.S.; Popov, V.A.; Popova, V.I.; Prokofiev, V.Z. 2000: A study of gas-liquid inclusions in beryl of the Mokrusha Mine, Middle Urals. **In** *Urals summer mineralogical school - 2000.* Ekaterinburg, Urals State Geological-Mining Academy. 21-22.

Anthony, J.W.; Bideaux, R.A.; Bladh, K.W. & Nichols, M.C. 1995: Handbook of Mineralogy, Volume II, Silica, silicates, part 1 and 2. Mineral Data Publishing, Tucson, Arizona. 904p.

Armbruster, T.; Libowitzky, E.; Diamond, L.; Auernhammer, M.; Bauerhansl, P.; Hoffmann, C.; Irran, A. Kurka, A. & Rosenstingl, H. 1995: Crystal chemistry and optics of bazzite from Furka Basistunnel (Switzerland). *Mineralogy and Petrology* **52**: 113-126.

Artini, E. 1915: Due minerali di Baveno contenenti terre rare: Weibyite e Bazzite. *Atti acad, Lincei* **5**: 24; 313-319

Bancroft, P. 1984: *Gem and Crystal Treasures.* Western Enterprises, Fallbrook, California, 488p.

Bartholomew, D.S. 1992: Base metal and industrial mineral deposits of Zimbabwe. *Geological Survey of Southern Rhodesia, Bulletin Mineral Resources Series No. 22*, Harare, Zimbabwe, 154 pages.

Bettencourt Dias, M. & Wilson, W.E. 2000: Famous mineral localities: the Alto Ligonha pegmatites, Mozambique. *Min Rec* **31**: 459-497.

Birch, W.D. & Henry, D.A. 1997: Gem minerals of Victoria. *Special Publication number 4, Mineralogical Society of Victoria* published in conjunction with the Royal Society of Victoria. 120p.

Blockley, J.G. & Meyers, J.S. 1990: Proterozoic rocks of the Western Australian Shield—geology and mineralisation. **In** Hughes, F.E. (ed.) *Geology of the Mineral Deposits of Australia and Papua New Guinea*, 607–615. The Australasian Institute of Mining and Metallurgy, Melbourne.

Breaks, F.W. & Janes, D.A. 1991: Granite-related Mineralization of the Dryden Area, Superior Province of Northwestern Ontario: *GAC-MAC Field Trip Guidebook B7*, Joint Annual Meeting, Toronto 1991.

Breaks, F.W. 1989: *Origin and evolution of peraluminous granite and rare-element pegmatite in the Dryden area, Superior province of northwestern Ontario*, Ph. D. thesis, Carleton University, Ottawa, Ontario, Canada.

Brown, G.E. & Mills, B. 1986: A High Temperature and Crystal Chemistry of Hydrous Alkali-rich Beryl from the Harding Pegmatite, Taos, Co. New Mexico. *Amer Min* **71**: 547-556.

Brown, J.A., 2004: Mineralogy and geochemistry of beryl and rare-metal bearing granitic pegmatites in the Kootenay Region of southestern British Columbia in *Geological Fieldwork 2003*, Geological Survey Branch, Ministry of Energy and Mines, Government of British Columbia. 167-184.

Burt, D.M. 1975: Beryllium mineral stabilities in the model system CaO-BeO-SiO_2-P_2O_5-F_2O-1 and the breakdown of beryl. *Econ Geol* **70**: 1279-1292.

Burt, D.M. 1978: Multisystem analysis of beryllium mineral stabilities: the system BeO-Al_2O_2-SiO_2-H_2O. *Amer Min* **63**: 664-676.

Černý, P. & Simpson, F.M. 1977: The Tanco pegmatite at Bernic Lake, Manitoba. IX. Beryl. *Can Min* **13**: 55-61.

Černý, P. 1992: Geochemical and petrogenetic features of mineralization in rare-element granitic pegmatites in the light of current research. *Applied Geochemistry* **7**: 393-416.

Černý, P. 2002: Mineralogy of beryllium in granitic pegmatites. **In**: E. S. Grew (ed.) Beryllium: Mineralogy, petrology, and geochemistry. *Rev Min Geo,* **50**: 691p.

Conklin, L.H. 1988: On Kunz and kunzite. *Min Rec* **18**: 369-372.

Conyers, L.B. & Goodman, D. 1997: *Ground-Penetrating Radar—An Introduction for Archaeologists*; AltaMira Press, Walnut Creek, California, 232p.

Cook, F.A. & Vasudevan, K. 1997: Georadar anisotropy in the Gotthard gneiss, Switzerland. *Geophysical Research Letters* **24**: 197-200.

Cook, F.A. 1997: Applications of Geophysics in Gemstone Exploration. *G&G,* **33**: 1; 4-23.

Cook, J.C. 1975: Radar transparencies of mine and tunnel rocks. *Geophysics* **40**: 5, 865-885.

Curran, J.M. 1897: The gems and precious stones of New South Wales. *Extract from the Proceedings of the Royal Society of New South Wales for 1896.* Angus and Robertson, Sydney, 73p.

Daniels, D.J. 1996: *Surface-Penetrating Radar.* The Institution of Electrical Engineers, London, 300p.

Della Ventura, G.; Rossi, P.; Parodi, G.C.; Montana, A.; Raudsepp, M. & Prencipe, M. 2000: Stoppaniite, $(Fe,Al,Mg)_4(Be_6Si_{12}O_{36}) \cdot (H_2O)_2(Na,\square)$, a new mineral of the beryl group from Latium (Italy). *European Journal of Mineralogy* **12**: 121-127.

Emlin, E.F. 1996: The Gem Belt of the Urals an Interminable Adventure, *World of Stones* **10**: 8-22.

Emsley, J. 2001: *Nature's Building Blocks: An A to Z Guide to the Elements.* Oxford University Press.

Feklichev, V.G. 1963: Chemical composition of minerals of the beryl group, character of isomorphism, and position of principal elements in the crystal structure. *Geochemistry* **4**: 410-421

Fersman, A.E. 1923: *Izumrudnye Kopi na Ural.* 82p.

Fersman, A.E. 1925: *Dragotsennye i Tsvetny Kamni SSSR* (Gemstones and colored stones of the USSR. Vol 2).

Fersman, A.E. 1929: Geochemische Migration der Elemente. III Smaragdgruben in Uralgebirge. Abh. z. praktischen Geologie u. Bergwirtschaft **1**: 74-116.

Fisher, J. 2002: The gem and rare-element pegmatites of Southern California. *Min Rec* **33**: 363-407.

Fisher, J.; Foord, E.E. & Bricker G.A. 1998: The geology, mineralogy, and history of the Himalaya mine, Mesa Grande, San Diego County, California. *R&M,* **73**(3), 156-180.

Foord, E.E. 1977: The Himalaya dike system, Mesa Grande district, San Diego County, California. *Min Rec* **8**: 461-474.

Foord, E.E.; Spaulding, L.B.; Mason, R.A. & Martin, R.F. 1989: Mineralogy and paragenesis of the Little Three mine pegmatites, Ramona district, San Diego County, California. *Min Rec* **20**, 101-127.

Gaines, R.V.; Skinner, H.C.W; Foord, E.E.; Mason, B., Rosenzweig, A. & King, V.T. 1997: *Dana's new mineralogy.* John Wiley and Sons. 1819p.

Garmo, T. 1995: *Norsk Steinbok.* 3. ed., Universitetsforlaget, Oslo. 300p.

Ginzburg, A.I. 1955: On the chemical composition of beryl. *Trudy Mineralogicheskoog muzeya Acad. Nauk SSSR* **7** (Transactions of the Mineralogical Museum, Academy of Sciences USSR).

Gordyenko V.V., Lyckberg, P. & Mikhailov V.V. 2000: Mineralogical and Geochemical features of the Högsbo Pegmatite. *Proceedings Russian Mineralog. Society* **4**.

Graindorge, J.M. 1974: A gemmological study of emerald from Poona, W. A. *The Austral Gem* **12**: 75–80.

Gramaccioli, C.M.; Diella, V.; Demartin, F.; Orlandi, P. & Campostrini, I. 2000: Cesian bazzite and thortveitite from Cuasso al Monte, Varese, Italy: a comparison with the material from Baveno and inferred origin. *Can Min* **38**: 1409-1418.

Hanley, J.B. 1951: Economic geology of the Rincon pegmatites, San Diego County, California. *California Division of Mines Special Report* **7B**, 24 p.

Hänni, H.A. 1978: Bazzit aus dem Tavetsch. *Lapis* **3**: 9; 29-30.

Hawthorne, F.C.; Cooper, M.A.; Simmons, W.B.; Falster, A.U.; Laurs, B.M.; Armbruster, T.; Rossman, G.R.; Peretti, A.; Günther, D. & Grobéty, B. 2004: Pezzottaite, $Cs(Be_2Li)Al_2Si_6O_{18}$, a spectacular new beryl-group mineral from the Sakavalana pegmatite, Fianarantsoa province, Madagascar. *Min Rec* **35**: 369-378.

Henley, H.F. & Brown, R.E. 2000: *Exploration data package. Clive 1: 100,000 Map Sheet Area. Vol. 1. Geology, mineral occurrences, exploration and geochemistry.* Geological Survey of New South Wales. GS1998/125 (unpublished).

Hochleitner, R. 2002: The emerald: mineralogically a beryl. **In**: *extraLapis English No.2: Emeralds of the World*, 10-15.

Huber, M. 2002: Beryll in Forschung und Technik. *extraLapis* **23**: 92-94.

Hutchinson, R.W. & Claus, R.J. 1956: Pegmatite deposits, Alto Ligonha, Portuguese East Africa. *Econ Geol* **51**: 757-780.

Jacobson, M. I. 1993: *Antero Aquamarines: Minerals from the Mount Antero—White Mountain region, Chaffee County, Colorado*. L. R. Ream Publishing, Coeur d'Alene, Idaho. 126p.

Jacobson, M.I., 1997: The Coolgardie pegmatite field, Western Australia. *Australian Journal of Mineralogy* **3**:2; 91–102.

Jacobson, M.I.; Calderwood, M.A. & Grguric, B.A. 2004: *A guidebook to the pegmatites of Western Australia.* (unpublished.)

Jahns, R.H. & Wright, L.A. 1951: Gem and lithium bearing pegmatites of the Pala District, San Diego County, California. *California Division of Mines Special Report* **7A**, 72 p.

Juve, G. & Bergstøl, S. 1990: Caesian bazzite in granite pegmatite in Tørdal, Telemark, Norway. *Mineral. Petrol.* **43**: 131-136.

Kalugin, P. 1888: Murzinka and Alabashka mines of colored stones of the Urals. *Zapiski Imperatorgskogo Mineralogicheskogo Obshchestva* **XXIV**, 254-260.

Kampf, A.R.; Gochenour, K. & Clanin, J. 2003: Tourmaline discovery at the Cryogenie mine, San Diego County, California. *R&M* **78**: 3; 156-168.

Kandutsch, G. 1998: Mineralien aus der Reisseckgruppe (Kärnten). *Lapis* **23**: 6; 27-28 & 41-45.

Keller, P.C. 1992: *Gemstones of East Africa.* Geoscience Press, Phoenix, AZ, USA, 144p.

Kokscharov N.I. 1988: Materialen zur Mineralogie Russlands. Bd. **10**. St-Petersburg: Gedruckt bei A. Jacobson

Kokscharov N.I. Materials on Mineralogy of Russia. St-Petersburg, Part 1, 1852-1855. Part 2, 1856, 243p (in Russian.)

Koshil, I.M., Vasilishin, I.S., Pavlishin, I.V. & Panchenko, V.I. 1991: Wolodarsk-Wolynskii: Geologischer Aufbau und Mineralogie der Pegmatite in Wolnyen, Ukraine. *Lapis* **16**: 28-40.

Kunz, G.F. 1905: Gems, jeweler's materials, and ornamental stones of California. *California State Mining Bureau Bulletin* **37**, 171p.

Lahti S. & Kinnunen K. 1993: A new gem beryl locality: Luumäki, Finland. *G&G* **29**: 1; 30-37.

Larson, W. 1972: The Queen reigns again. *Lap J* **26**: 1002-1032.

Larson, W. 1977: The best of San Diego County. *Min Rec* **8**: 507-515.

Laskovenkov, A. 1992: Unpublished report on the Mariinski Emerald Mine.

Laurs, B.M.; Simmons, W.B.; Rossman, G.R.; Quinn, E.P.; McClure, S.F; Peretti, A.; Armbruster, T.; Hawthorne, F.C.; Falster, A.U.; Günther, D.; Cooper, M.A. & Grobéty, B. 2003: Pezzottaite from Ambatovita, Madagascar: a new gem mineral. *G&G* **39**: 284-301.

Lazarenko, E.K.; Pavlishin, V.I.; Latisch V.T. & Sorokin J.G. 1973: *Mineralogija Igenezis Kamernich Pegmatitov Volini.*

Lees, B.K., 1998: The Application of Ground-Penetrating Radar to Mineral Specimen Mining. *Min Rec* **29**: 2; 50 & 145-153.

Legun, A. 2004: The potential for emeralds in B.C.—A preliminary overview in *Geological Fieldwork 2003,* British Columbia Ministry of Energy and Mines, Paper 2004-1, 219-229.

London, D. & Evensen, J.M. 2002: Beryllium in silicic magmas and the origin of beryl-bearing pegmatites. **In**: Grew, E.S. (ed.) Beryllium: Mineralogy, Petrology and Geochemistry: *Rev Min Geo* **50**, 445-486.

Lyckberg, P. & Hawthorne, F.C. 1997: Foitite from the Kazionnitsa pegmatite, Alabashka pegmatite Field, Ural Mountains, Russia. *Tourmaline 1997.* Moravian Museum, Brno, Czeck Republic and University of Manitoba, Winnipeg, Canada.

Lyckberg, P. & Zagorsky, V.Y., 2000: Gem pegmatites in the Ural Mountains, Russia (abstracts.) *Min Rec* **31**: 2, 183.

Lyckberg, P. 1975: Västsveriges Pegmatiter. Unpubl. Report *Göteborgs Högre Samskola.*

Lyckberg, P. 1981: Pegmatiter I Göteborgstrakten, speciellt Högsbo Pegmatitbrott, *Hvitfeldtska Gymnasiet,* Göteborg. (unpubl.)

Lyckberg, P. 1992: Update on the Mursinka-Alabsahka Gem Pegmatite District of the Urals. (unpubl.)

Lyckberg, P. 1993: Eine Reise zum bekannten "Mursinka-Alabashka Pegmatit und dem Amethystgebiet NNE von Jekaterinburg im Ural (2). *MW* **4**: 2; 51-56.

Lyckberg, P. 1995: Mursinka ädelstenarnas hemvist på jorden. *Stenoliten* **1**: 2; 22-25 & 43.

Lyckberg, P. 1995: Topaz and beryl-bearing gem pegmatites of the Alabashka-Mursinka-Adui district in the Ural Mountains, Russia (abstract.) *Min Rec* **26**: 1; 67-69.

Lyckberg, P. 1997: Giant Tourmalines in Pegmatite. *Tourmaline Symposium, Nove Mesto na Morave, Czeck Republic.*

Lyckberg, P. 1997: Prediction of Gem Pockets in Ural Mountain pegmatites, Russia (abstract.) *R&M* **72**: 3; 187

Lyckberg, P. 1997: Some notes of the fractionation in pegmatites in and around the Bohus-Iddefjord granite in western Sweden and SE Norway. *International Symposium of Miarolitic Pegmatites,* Milano, Italy.

Lyckberg, P. 1999: A recent study of the miarolitic gem pegmatite fields of the Mursinka-Adui intrusions of the central Ural mountains, Russia and a comparacy to G. Flinks travels in 1889. *Flink Symposium 1999 Swedish Min Soc.* Swedish Museum of Natural History,

Lyckberg, P. 2001: Gem Pegmatite and Greisen Deposits of Russia During the Twentieth Century, *Min Rec* **32**: 1; 44-45.

Lyckberg, P. 2004: Recent gem beryl production in Finland. *G&G* **40**: 256-258.

Mandarino J.A. & Malcolm E. B. 2004: Fleischer's Glossary of Mineral Species. 309p. The Mineralogical Record, Tucson, AZ.

Marshall, D.; Groat, L.A.; Giuliani, G.; Murphy, D.; Mattey, D.; Ercit, T.S.; Wise, M.A.; Wengzynowski, W. & Eaton, W.D. 2003: Pressure, temperature, and fluid inclusion conditions during emerald precipitation, southeastern Yukon, Canada: fluid inclusion and stable isotope evidence: *Chemical Geology*, **194**: 287-199.

Meagher, E. P. & Gibbs. G.V. 1977: The polymorphism of cordierite: II. The crystal structure of indialite. *Can Min* **15**: 43-49.

Mulligan, R. 1968: *Economic Geology Report #23—Geology of Canadian Beryllium Deposits.* Geological Survey of Canada, Ottawa, Department of Energy Mines and Resources, 109p.

MINERALIENTAGE
Europe's Top Show
MÜNCHEN
2005
OCT. 28 - 30
GEOFA: October 28, '05
(Professionals)
DMF
recommended
Special Exhibit
AGATE
Beyond belief!
Wonders are collectable.
www.mineralientage.com

Murphy, D.C.; Lipovsky, P.S.; Stuart, A.; Fonseca, A. & Piercey, S.J. 2002: What About Those Emeralds, Eh? *Poster at www.geology.gov.yk.ca/metallogeny/index.html*

Nassua, K.; Prescott, B.E. & Wood, D.L., 1976: The deep blue maxixe-type colour center in beryl: *Amer Min* **61**: 100-107.

Neufeld, H.L.D.; Groat, L.A.; Mortensen, J.K. & Israel, S. 2004: Origin of the Regal Ridge emerald property, southern Yukon: *Can Min* submitted.

Olliver, J.G. & Stevenson, B.G. 1984: Pegmatites in the Olary Province—feldspar and beryl mining. *Mineral Resources Reviews South Australia* **154**: 22–35.

Ossovski, G. 1867: Geologisch-geognostischer Abriss des Volynsker Gouvernements. *Arbeiten des Volynsker statistischen Komitees für 1867*, Shitomir.

Palmer, A., 1990: Poona, W.A and the seekers of its emeralds *Lapidary Industries*. North Lake, WA, 97p.

Peretti, A.; Armbruster, T.; Günther, D.; Grobéty, B.; Hawthorne, F.C.; Cooper, M.A.; Simmons, W.B.; Falster, A.U.; Rossman, G.R. & Laurs, B.M. (2004). The challenge of the identification of a new mineral species: example "pezzottaite". *Contributions to Gemology* **3**: 1-12.

Petterd, W. F. 1910: The minerals of Tasmania. *Papers and Proceedings of the Royal Society of Tasmania*. 1–221.

Pezzotta, F.P. 2001: Madagascar— A Mineral and Gemstone Paradise. *extraLapis English No.1*. 97p.

Platonov, A.N.; Taran, M.N.; Klyakhin, V.A. 1989: On two colour types of Mn^{3+}-bearing beryls. *ZDGG* **38/**4: 147-154.

Platonov, A.N.; Taran, M.N.; Minko, O.E. & Polshyn, E.V. 1979: Optical absorption spectra and nature of colour of iron-containing beryls. *Izvestiya Akademii Nauk SSSR, Seriya Geologicheskaya* **10**: 54-68.

Popova V.I., Popov V.A., Poljakov V.O. Tcherbakova E.P. 1982: *Pegmatit Ilmenskich Gor*. - Sverdlovsk: UNC AN CCCP. 48p.

Popova, V.I.; Popov, V.A. & Kanonerov, A.A. (2002) Murzinka: Alabashka Pegmatite Field. *Min Alm* **5**. 128p. Moskow, Russia.

Proctor, K. 1984: Gem pegmatites of Minas Gerais, Brazil: exploration, occurrence, and aquamarine deposits: *G&G* **20**: 4; 78-100.

Robinson, S. 1825: *Catalogue of American Minerals with their Localities*. Cummings, Hilliard & Co. Boston. 316p

Rohtert, W.R.; Quinn, E.P.; Groat, L.A. & Rossman, G.R. 2003: Blue beryl discovery in Canada: *G&G* **39**: 4 327-329.

Rose, G. 1837-1840: *Mineralogische geognostische Reise nach dem Ural, dem Altai und dem Kaspischen Meer*. Berlin.

Ruegg, H.-R. & Klemm, L. 2002: Silbererze aus dem Bergwerk Goppenstein, Wallis (Schweiz). *Lapis* **27**: 3; 25-34.

Rynerson, F.J. 1967: *Exploring and Mining for Gems and Gold in the West*. Naturegraph, Healdsburg, California, 204 p.

Satterly, J. 1941: Geology of the Dryden-Wabigoon Area: *Ontario Department of Mines 50th Annual Report* **1**: 2; 67p.

Schaller, W.T. & Fairchild, J.G. 1932: Bavenite, a beryllium mineral pseudomorphous after beryl. *Amer Min* **17**: 409-422.

Schebasta, K. & Weninger, H. 1978: Kleines aus dem Grossen Fleisstal. *Lapis* **3**: 12; 26-30.

Schmetzer, K.; Berdesinski, W. & Bank, H. 1974: Über die Mineralart Beryll und Absorptionsspektren. *ZDGG* **23**: 1; 5-39.

Schwarz, D., 1991: Australian emeralds. *The Austral Gem* **17**: 12; 488–497.

Schwarz, D.; Guiliani, G.; Grundmann, G. & Glas, M. 2002: The Origin of Emerald In: *extraLapis English No.2: Emeralds of the World*; 10-15.

Shigley, J. E.; Thompson, T.J & Keith. J.D. 2003: Red beryl from Utah: a review and update. *G&G* **39**: 302-313.

Simmons, W. B.; Falster, A.U.; McClure, S.F.; Quinn, E.P.; Rossman, G.F.; Hawthorne, F.C. & Laurs, B.M. 2003: A new saturated purplish pink Cs-"beryl" from Madagascar: preliminary analyses. G&G **39**: 50-54.

Simmons, W.B.; Webber. K.L.; Falster, A.U. & Nizamoff J.W. 2003: *Pegmatology—Pegmatite mineralogy, petrology, and petrogenesis*. Rubellite Press, New Orleans, Louisiana. 176 p.

Sinkankas, J. 1959: *Gemstones of North America* **1**. Van Nostrand, New York, 675 p.

Sinkankas, J. 1976: *Gemstones of North America* **2**. Van Nostrand, New York, 494 p.

Sinkankas, J. 1981: *Emerald and other Beryls*. Van Nostrand, New York, 665p.

Sinkankas, J. 1997: *Gemstones of North America* **3**. Geoscience Press, Tucson, 526 p.

Skrigitill, A.M. 1996: Gemstones in the pegmatites of the Eastern Pamir. *World of Stones* **11**: 16-25.

Soustzinski P.P. 1925: A scetch of the localities of colored stones in south-east Transbaikal District (in Russian), *Trans-actions of the Institute of Economic Mineralogy and Metallurgy* („*Lithogaea*") **16**/88 Scientific Department of the Supreme Council of National Economy, Moscow, USSR.

Sutherland, L. 1991: *Gemstones of the Southern Continents*. Reed Books, Sydney, 256p.

Svirina E.S. 1994: Earl Perovsky and his beryl collection in the mining museum. *World of Stones* **3**; 35-38.

Talantsev A.S. 1988: *Ural chamber pegmatites*. Moscow, Nauka, 144p, (in Russian).

Thompson, W.B.; Joyner, D.L.; Woodman, R.G. & King, V.T. 1998: *A collector's guide to Maine mineral localities*. Maine Geol. Survey, Natural Resource Information and Mapping Center, Department of Conservation. 106 p.

U.S. Geological Survey Minerals Yearbook, 2002, Beryllium

Vlasov, K.A. & Kutakova, E.I. 1960: *Izumrudnye Kopi*, Moscow, Academiza Nauk USSR, 252p.

Von Knorring, O. 1970: Mineralogical and geochemical aspects of pegmatites from orogenic belts of equatorial and southern Africa. In: Clifford, T.N. and Gass, I.G. (eds.) *African Magmatism and Tectonics*, Oliver & Boyd, Edinburgh, 157-184.

Walton, L. 2004; Exploration Criteria for Coloured Gemstone Deposits in the Yukon: *Yukon Geol. Survey*, 184p.

Webb, G. & Sutherland, L. 2002: *Gemstones of New England, New South Wales*. The Mineralogical Society of New South Wales. 25 pp.

Weber, F.H. 1963: Geology and mineral resources of San Diego County, California. *California Division of Mines and Geology County Report* **3**, 309 p.

Weiss, S. 1994: Blauer Kluftberyll aus der Südschweiz. *Lapis* **19**: 6; 25-40.

Wild, G.O. & Klemm, R. 1926: Mitteilungen über spektroskopische Untersuchungen an Mineralien, Smaragd, *Zentralblatt für Mineralogie Abt. A* **1**: 21-22.

Wilson, B. 1999: Gemstone occurrences in British Columbia: *Canadian Rockhound* **3**: 1; 1-15.

Wood D.L. & Nassau, K. 1968: The characterization of beryl and emerald by visible and infrared absorption spectroscopy. *Amer Min* **53**: 777-800.

Zagorski, V.E.; Peretyazhko, I.S & Shmakin, B.M. 1999: *Granitnie Pegmatit Tom 3, Miarolnij Pegmatit, Novosibirsk.* Nauka. Sibirskaja Isdatelskaja Firma RAN.

Websites

Guide to Copper Beryllium, by Brush Wellman, no author credited: http://www.brushwellman.com/alloy/tech_lit/GuideToCopperBeryllium.pdf

James Webb Space Telescope: http://www.jwst.nasa.gov/Telescope/mirrortale/

USGS: Mineral Commodity Summaries, January 2004, Beryllium: http://minerals.usgs.gov/minerals/pubs/commodity/beryllium/

Abreviations: ***Amer Min***: *American Mineralogist;* ***Austral Gem***: *Australian Gemmologist;* ***Can Min***: *Canadian Mineralogist;* ***Chem Geol***: *Chemical Geology;* ***Econ Geol***: *Economic Geology;* ***G&G*** *Gems & Gemology;* ***Lap J***: *Lapidary Journal;* ***Min Alm***: *Mineralogical Almanac;* ***Min Rec***: *The Mineralogical Record;* ***MW*** *Mineralien Welt;* ***R&M***: *Rocks & Minerals;* ***Rev Min Geo***: *Reviews in Mineralogy and Geochemistry*

Author's Adresses

Dmitriy Belakovskiy
Fersman Museum Moscow
Leninskiy Prospekt 18-2; 117071 Moscow; Russia
dmvrz@aol.com

Dr. Bruce Cairncross
Department of Geology; University of Johannesburg
P.O. Box 524, Auckland Park, 2006; Gauteng, South Africa
bc@na.rau.ac.za

Alexander U. Falster
Department of Geology & Geophysics ;
University New Orleans
2000 Lakeshore Drive, New Orleans, LA 70148; USA
afalster@uno.edu

Jesse Fisher
380 Pennsylvania Ave.; San Francisco, CA 94107; USA
Jef520@aol.com

Dr. Lee A. Groat
Department of Earth and Ocean Sciences
University of British Columbia
Vancouver, BC Canada V6T 1Z4
lgroat@eos.ubc.ca

Michael Huber
c/o Christian Weise Verlag
Orleansstrasse 69; D-81667 München; Germany
lapis@lapis.de

Dermot Henry
Geosciences Museum Victoria
GPO Box 666E; Melbourne, Victoria Australia 3001
dhenry@museum.vic.gov.au

Dr. Rupert Hochleitner
Mineralogische Staatssammlung
Theresienstrasse 41; D-80333 München; Germany
rh.minstaatsslg@lrz.uni-muenchen.de

Mark Ivan Jacobson
108 Farmington Drive; Lafayette, LA 70503; USA
MarkIJacobson@chevrontexaco.com

Dr. Bruce Jarnot
1750 Twirly Court; Prince Frederick, MD 20678; USA
bmjarnot@aol.com

Dr. Guanghua Liu
AAA Minerals International
Französische Allee 24; 72072 Tübingen; Germany
ghliu@aaamineral.com

Peter Lyckberg
BP 2785; L-1027 Luxembourg; Luxembourg
lyckberg@pt.lu

Luiz A. Menenez
R. Esmeralda, 534
30410-080 Belo Horizonte, Minas Gerais; Brazil
lmenezesminerals@uol.com.br

Alfredo Petrov
531 North James Street; Peekskill, NY 10566; USA
alfredopetrov@earthlink.net

Tony L. Potucek
3701 Country Club Drive; Show Low, AZ 85901; USA
Tony_Potucek@abitibiconsolidated.com

Dr. William "Skip" Simmons
Department of Geology & Geophysics;
University New Orleans
2000 Lakeshore Drive, New Orleans, LA 70148; USA
wsimmons@uno.edu

Dr. Fabio Tamagnini
Via Franco Antonini 7; 00053 Civitavecchia; Italy
tamagninifabio@interfree.it

Lee Van Iderstine
348 Green Rock; Shelton, CT 06484; USA
N1SCQ@aol.com

Dr. Stefan Weiss
c/o Christian Weise Verlag
Orleansstrasse 69; D-81667 München; Germany
lapis@lapis.de

Dr. Michael Wise
Department of Mineral Sciences, National Museum of Natural History Smithsonian Institution
Washington, DC 20560; USA
wise.michael@nmnh.si.edu